The Genes That Make Us

Professor Edwin Kirk is both a clinical geneticist and a genetic pathologist, a rare combination. As a clinician, he sees patients at Sydney Children's Hospital, where he has worked for more than 20 years; his laboratory practice is in the New South Wales Health Pathology Genomics Laboratory at Randwick.

Kirk is a conjoint appointee in the School of Women's and Children's Health at the University of New South Wales, an experienced medical educator, and currently Chief Examiner in Genetics for the Royal College of Pathologists of Australasia. He is also a respected researcher, working in the fields of cardiac genetics, metabolic diseases, and intellectual disability, as well as studying reproductive carrier screening, and is a co-author of more than 100 publications in scientific journals, which have been cited by other researchers more than 4,000 times. He is one of the co-leads and public faces of the $20 million Mackenzie's Mission carrier screening project.

Kirk lives in Sydney with his wife and three children. In his spare time, he competes in ocean swimming races, slowly, and plays the saxophone, loudly.

The Genes That Make Us

human stories from a revolution in medicine

Edwin Kirk

SCRIBE

Melbourne • London

Scribe Publications
18–20 Edward St, Brunswick, Victoria 3056, Australia
2 John Street, Clerkenwell, London, WC1N 2ES, United Kingdom

First published by Scribe 2020

The information in this book is general in nature and should not
be considered to be personal medical advice. Readers are advised
to contact their own doctors or other health professionals in
relation to any medical concerns regarding their own or their
children's health, and should seek medical advice in relation to
pregnancy-related issues, including screening and other tests
during or before pregnancy.

Typeset in 11.5/15 pt Sabon by the publishers

Printed and bound in Australia by Griffin Press, part of Ovato

Scribe Publications is committed to the sustainable use of natural
resources and the use of paper products made responsibly from
those resources.

9781925849394 (Australian edition)
9781912854363 (UK edition)
9781925938425 (ebook)

Catalogue records for this book are available from the National
Library of Australia and the British Library.

scribepublications.com.au
scribepublications.co.uk

To my parents, Robin Enfield Kirk and Rosalie Saxby.
With much love and thanks, for nature and nurture.

Author's note

This book contains numerous descriptions of patients I have seen. In order to protect patient confidentiality, the descriptions have been extensively altered, sometimes by combining events that happened to several people. Consent has been provided as appropriate. Where the stories are based on real events, my intention is that it should be impossible to identify any particular individual from their descriptions here. An exception is that, if a patient's story has previously been published in the medical literature, I have generally kept the key elements from the published version. I also tell the stories of some people who were not my patients, including Jesse Gelsinger and Mackenzie Casella. For both of the latter, extensive media coverage has already occurred. Mackenzie's parents have read and approved the account of her life in chapter 11.

Contents

Preface

An end, and a beginning

Genetics has taken me to some unexpected places. A basement stacked with hundreds of boxes of mice. A mosque in Pakistan, and another in the suburbs of Sydney. A ballroom filled with hundreds of people, every one of them seated in front of two small glasses of poison.

Mostly, though, there's nothing about the life of a geneticist that would strike a casual observer as exotic. Our days are filled with meetings and paperwork. We see patients in clinic rooms, or on the wards, like any doctor. Our labs have as much room devoted to generic-looking office space as they do to high-tech machinery. And even the high-tech machinery doesn't look like much. There's an occasional device with futuristic flair, but much of our equipment sits squarely in the 'boring grey box' school of industrial design.

Yet you shouldn't let yourself be fooled by outward appearances. Remarkable things are happening in genetics, a quiet revolution that has already dramatically changed some parts of medicine, and is coming for the rest. Within the next few years, having your genome sequenced will become routine. There's a good chance you'll have yours done one of these days, if you haven't already. A decade or two from now, your family doctor will have your genetic information on file, as much a part of your record as your blood pressure, your weight, and the medications you take.

There's a standard job-interview question: 'where do you see yourself in ten years from now?' When this question gets asked of a clinical geneticist, the possibility often comes up that, in ten years from now, this will be a dying specialty — not because genetics will become less important, but because it will be *so* important that every doctor will need to have mastery of the field, and nobody will need a doctor who just does genetics. I've been hearing predictions like this for nearly a quarter of a century, but they have never seemed further from coming true than they are today. Instead, a handful of specialists — neurologists particularly, but some cardiologists, endocrinologists, and others — have embraced genetics, while most doctors have been way too busy with all the advances in their own fields to even try to keep up with ours. Meanwhile, our numbers have grown steadily, but from a tiny base, so that we are still relatively obscure. Even other doctors are often rather vague about what a geneticist does.

So what *do* we do? Unusually for medical specialists, our patients are not limited to one age group, or to people with problems affecting a particular organ. Sometimes, we are involved in peoples' lives before they are even conceived; sometimes, when they are in the womb. We see babies, and children, and adults who are hoping to have children. We see grandparents because they have developed a genetic disorder late in life, or because a faulty gene is being tracked through a family to find people who are at risk. Sometimes, the first time a person has a genetic test is after they have died. My colleague David Mowat talks about the scope of our job being to provide care not just 'from womb to tomb', like a general practitioner, but 'from sperm to worm'.

The thing that links all of our patients is, of course, genes — and genetic disease in particular. The questions we try to answer are fundamental ones. How can we have a healthy baby? What caused my child's heart condition? Will I develop Huntington disease, like my father and his father?

In asking questions like these, people are letting us into their lives, often at times of strong emotions, often when there is loss and grief. Over my career so far, I've been privileged to share part of the lives of thousands of people. Luckily for me, this time has also been an unprecedented period of growth in our understanding of genetics, a time of ever-accelerating discovery.

For me, it began in the mid-1990s. I was a junior doctor then, working in the intensive care unit of a children's hospital in Sydney. One of my patients was a tiny baby, born with congenital heart disease, smaller than she should have been, and seemingly unwilling to breathe on her own. Machines were keeping her alive.

The results of genetic testing had come back, and a meeting was arranged so that one of the hospital's clinical geneticists could explain the result to the parents. Someone from the intensive care unit would usually sit in on such meetings, and so by chance it happened that I was there, witness as a young mother and father received the worst news of their lives.

The clinical geneticist was Dr Anne Turner. Later, she would become a colleague, and one of my closest friends. Anne is witty, kind, loyal, an inveterate traveller and a bon vivant, a loving mother, and now a besotted grandmother. Back then, I knew her only as my senior in the medical hierarchy, a respected specialist in a small and somewhat obscure field of medicine.

Anne does not remember our first meeting as I do, because her focus was on the difficult task she had come to do. She was there to tell those parents that their daughter was going to die.

Testing had shown that the little girl had an extra copy of chromosome 13, a condition known as Patau syndrome. Affected babies are small, and often have heart conditions and brain malformations or other physical problems. Their brains do not function normally, even to the point that they cannot control the body's breathing normally. Almost all babies with Patau syndrome die within the first year of life, mostly within a few days or weeks of birth. The rare survivors have severe intellectual disability and other health problems.

This diagnosis explained all the problems we had been

struggling to treat — and, particularly in a baby who could not breathe by herself, it meant the prognosis was grim.

Giving bad news is hard. The shock and grief parents feel when they hear bad news about their children is intense, difficult to bear for those in the room with them. It puts particular pressure on the person delivering that news, in part because it is difficult to escape the feeling that you are the cause of the pain. Sometimes, you have had time to get to know and like the people you are hurting, but, even when it's someone you've only just met ... it's tough.

So why did being present at such an occasion draw me towards a career in genetics, rather than pushing me away? Mostly, I think, it was the way that Anne approached the task. Her warmth and gentleness tempered a direct, sure manner. She explained what a chromosome was, what it meant to have an extra chromosome, and what that meant for the little girl. One of the parents asked if it wasn't possible to fix the problem, to remove the extra 13. Patiently, Anne explained that the problem was deep inside every cell of the child's body, that it had been that way since conception, and that there was no way to undo this. She listened when it was time to listen; she spoke when it was time to speak. She acknowledged the love the couple had for their daughter, and the pain they were feeling. But she gave no false hope. What Anne showed me on that day was a way to practise medicine in a place where the most advanced science and the deeply human meet.

My goal in this book is to reveal the humanity in human genetics, through the stories of those whose lives are most affected by it. If you've come for the science, keep reading — genetics is by far the most exciting of the modern sciences, and there's plenty of science in these pages.

But the story of human genetics is, above all, a story of *people*. It is the story of the people whose lives are affected by genetics — which is everyone, really, but some far more obviously, and some far more harshly than others. It is the story of that tiny baby, doomed from the moment she was conceived. It is the story of the scientist in the lab, looking down her microscope and reading the news of the little girl's fate, written in the language of the cells. It is Anne's story, and mine. It is the story of the people who first learned what a chromosome was, and how it might link to disease.

Most of all, perhaps, it is the story of two young parents, grieving and bereft — but armed with knowledge and understanding, allowing them to face the future.

I

Easier than you think

Professor Kirk makes genetics as easy as A C G T.

Seamus Kirk[1]

My friend and colleague Steve Withers, a geneticist himself, often refers to others as having 'a brain the size of a planet'. Many people think that you need a bulging cranium to understand genetics. There's an aura of difficulty around the subject … which turns out to be a complete con. Genetics is remarkably straightforward. If, by the end of high school, you could manage primary-school mathematics with reasonable confidence, you will have no difficulty with the essentials of genetics.

Why do people think it's hard? Perhaps it's just that there is a great deal of detail — thousands of different conditions, all of which vary in their severity, many of which overlap with each other. To fully understand genetic disease, you need to know a bit about how cells work, and there is an awful lot of detail there, too. It's all just information piled on information, though — anyone can understand any individual part of it.

To prove the point: perhaps the most important piece of information in genetics is the relationship between DNA and proteins. This relationship is similar to, but *much simpler than*,

---

1   It's possible that my son is not an entirely impartial critic.

7

the relationship between letters and words. Here are the facts:

Proteins form a lot of the 'stuff' of the body — they are the building blocks of cells, and of the padding between the cells. Anytime your body has a job to do, it gives it to a protein. If your cells wanted to make a car, every single mechanical and electrical component would be made from proteins ... and so would the garage you parked the car in; it's not just for moving parts. Proteins themselves are made up of amino acids.

DNA is a chemical that contains information. This information is written in an alphabet with only four letters: A, C, G, T. They stand for four nucleobases,[2] the chemical building blocks of DNA.

Unlike English, the language of DNA has only 21 words. The spelling for those words always involves three nucleobases — it's a code of threes. In English, CAT means a furry parasite, but, in this language, it means the amino acid histidine. There are 20 amino acids represented in this language, and the 21st word is 'stop'. A gene is a stretch of DNA that codes for a particular protein — so it's a string of groups of three that say, 'Put a histidine in. Then put a glycine in. Then a proline. Okay, now stop.'

You can think of nucleobases as the letters, amino acid names as the words they spell, and genes as sentences. Each sentence explains how to build a particular protein, and each molecule of DNA contains many of these sentences. It's a manual for building parts of the body.

That's it. The fundamental basis of genetics. Far simpler than learning to read, and we ask six-year-old children to do that. Even better, there's no need to actually learn the language — you just need to understand that there *is* a language, and how it works. After more than 20 years in genetics, I only know the spelling of three or four of the words in the code. The rest I look up when I need to.

There are no concepts in genetics more complex than the

---

2   You may be more familiar with the term 'nucleotide', which is a nucleobase attached to the other structural elements of DNA.

one you just learned, if you didn't know it already. The rest is just detail.

Fortunately, although genetics is simple, it is also fascinating. Take chromosomes, for instance.

Chromosomes, the physical form our DNA takes within cells, are wholly remarkable structures. You've probably seen pictures before, but, just in case, here's an example.

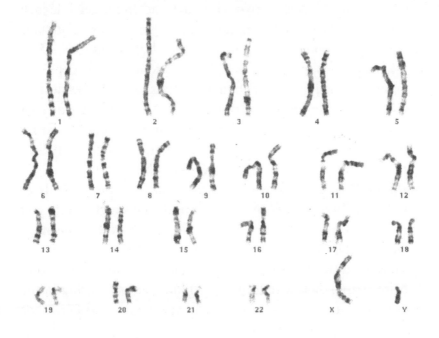

This is a particularly good set of chromosomes — they're mine. One of the lesser known perks of training in genetics used to be the chance to prepare and examine your own chromosomes, and who could resist an opportunity like that? Today's trainees don't get the chance, for fear they might find out something they don't want to know. It's a pity — there is something immensely satisfying about staring at your own genome down a microscope. I imagine it's a bit like seeing video of your own heart after an operation, but without the inconvenience of having your chest cracked open to get the pictures.

A genome is the totality of an organism's genetic information, and every living thing has one — you, me, a slug, a blue whale, the kale in the salad you had for lunch, the microbes living under your waiter's fingernails.[3] Bacteria have genomes; protozoa and fungi have them; viruses have them, too. And in everything from bacteria on up, the genome is organised into chromosomes. The number of chromosomes varies enormously between species, and there is no clear link between how complex an organism is and how many chromosomes it has. Bacteria, to be sure, have only one or two, compact and circular. Male jack jumper ants — far more complex than a bacterium — also have only a single chromosome. But Atlas blue butterflies have 450.

The chromosomes you see in the picture were captured at a very particular moment in their existence. It's easiest to look at chromosomes when they are like this, at a point part way through cell division. They are compressed, and easily recognised as separate structures. Humans (mostly) have 23 pairs of chromosomes. They are 46 long, thin threads of DNA, totalling about two metres, in each of the trillions of cells in your body. Two metres may not sound like much, until you remember that a typical cell nucleus — which holds almost all of the cell's DNA — is only six *millionths* of a metre across. If the nucleus were the size of your lounge room, and DNA were made of string, there would be 1,000 kilometres of string in the room with you — enough to stretch from London to Berlin, or from San Francisco to Portland.

Most of the time, that string is not bunched up into the tight bundles that you're seeing in the picture. It's a delicate gossamer, stretched and twirled through the nucleus, not completely on the loose but organised, coiled around proteins called histones. This DNA-protein combination is called chromatin, and it is the stuff of life.

DNA, famously, carries information. It carries it through generations, and through deep time. Your DNA is the result of a continuous, unbroken chain of events that has lasted for billions of years. It has been copied, over and over, subtly changing as it

---

3   Eww.

went, starting with the first, simple living things that emerged in some warm, shallow, long-forgotten sea. It has endured through many different forms, through mammals, through proto-humans, through the whole of humanity's existence, until your conception. It carries the memory of that long journey with it. Though we may forget, our genes do not.

Work in genetics for a while, and each chromosome develops its own flavour — not a personality exactly; it's more that there are things that spring to mind when someone mentions each of them. Chromosome 1 has a pale section near the top — you can see it easily on mine. Remove that section on one of the two copies at conception, and the child who results will have intellectual disability, and a distinctive facial appearance, with deep-set eyes and low-set ears. Chromosome 7 is home to the cystic fibrosis gene, the goal of a race to discovery and a rich scientific prize (that race was won by Lap-Chee Tsui, a dual Hong Kong/Canadian citizen who was working in Toronto at the time). 17 is where *BRCA1*, one of the breast cancer genes, can be found. The story of the race to find *BRCA1* is a darker one, and its consequences are still playing out in the patent courts and in people's lives today. Chromosome 15 is associated with Prader-Willi syndrome and Angelman syndrome, two very different disorders forever locked together, strange dance partners. There are a few places in the human genome where genes remember which parent they came from and are switched on or off accordingly, and chromosome 15 contains one such region. Chromosomes 13, 14, 15, 21, and 22 are the acrocentric chromosomes: their waists are where their heads should be. Sometimes, they actually fuse together, head to head (a Robertsonian translocation). The Y is a wasteland, a dying chromosome littered with the corpses of broken genes. It hardly has any reason left to exist, and yet it struggles on.

Chromosome analysis, also known as karyotyping, was the original genetic test. There were other medical tests before the karyotype that could detect genetic disorders — examination of a film of blood under a microscope to diagnose sickle cell disease, for example. But this was a test that was *purely* genetic. More than that, it was the first, and, for a long time, the only, *genomic* test: it examines the whole of a person's genome for abnormalities in one go. It's a bird's-eye view, lacking in detail by today's standards, but it has stood the test of time, and we are still using it today.

It has always intrigued me the way human experience builds up around a new technology. Take flying, for instance. Hardly any time had passed after the invention of powered flight before aviation developed its own received wisdom. Aviate, navigate, communicate.[4] There are old pilots, and there are bold pilots, but there are no old, bold pilots. Nothing is less use to a pilot than altitude above you and runway behind you.

The same thing has happened with cytogenetics (the study of chromosomes), and even with the newer genetic technologies. There are known traps for young players. There's the way we've always done things (it's always worked, so why change it?). And, already, young though the field is, there is tradition.

Part of that tradition has to do with the naming of parts. Look at the chromosomes in the picture and you'll see that some have a waist part way along their length. This is the centromere, the structure that anchors and guides the chromosome during cell division. It's never exactly in the middle of the chromosome, which means that there is a short arm and a long arm on either side of it: these are named the p and q arms.

Why p and q? In 1966, early in the story of chromosome analysis, a meeting was convened in Chicago[5] to discuss stan-

---

4   For when a pilot is in difficulty: aviate — first, do what you need to in order to keep the aircraft flying; navigate — next priority is to figure out where you are and where you might be able to land; communicate — once the first two are under control, you need to talk to the ground, and to other aircraft.

5   Perhaps because of the outcome, there's a myth among geneticists that this happened at the Paris nomenclature meeting of 1971. When you read the records of that meeting, however, it's obvious that the p/q question had long been settled by then. The same story alleges that q was chosen because it's the next letter in the alphabet. I have been telling medical students this tale for many years, and never bothered to check until I was writing this. My apologies to all those I've misled.

dardisation in the description of chromosomes. It was decided that the short arm would be the p arm — for 'petit', French for 'small'. There had been discussion of calling it s for 'short', but the French cytogeneticist Jérôme Lejeune was evidently a persuasive man. And perhaps this was a tactical concession by those who wanted to claim the long arm for themselves.

By the time p was agreed upon, it was late in the night. English speakers pushed for the long arm to be l, but it was pointed out that this could easily be confused with the numeral 1. Nobody wanted to let the French have both arms, so there was something of a stalemate. This was broken by the English geneticist Lionel Penrose, who suggested q, because it favoured no language, and because, in another branch of genetics, there was a famous equation, $p+q=1$, which suggested that with the p arm and the q arm, you have the whole of the chromosome. At this point, it seems, everyone was sick of the dispute, and welcomed the chance to settle the matter and get to bed.

Looking along the arms of a chromosome, cytogeneticists learned to recognise patterns of light and dark staining, due to the interaction of the chromosome material with the dyes used in preparing the slides. You can see these bands in my chromosomes. We've already looked at the top (the end of the p arm) of chromosome 1; combine that with the fact that 1 is the largest chromosome and you won't have trouble finding it. Now look at chromosome 7 — it's a medium-sized chromosome with a prominent dark band near the end of the p arm. You'll never mistake a 1 for a 7 now, and you should be able to pick out either in a crowd. Congratulations! You're on your way to becoming a cytogeneticist.

Sorting the chromosomes by size, from 1 to 22 plus the X and Y (although it turned out that 21 is actually a little smaller than 22), and by their bands, with ever finer divisions in those bands, led to a system of addresses. Chromosome 1 was divided into 1p and 1q. 1p was divided into 1p1, 1p2, 1p3 ... and so on, until today we have addresses like 1p36.33 — chromosome number, chromosome arm, band (3), sub-band (6), sub-sub-band

(3), and even sub-sub-sub-band (3). Each of these is only visible at finer and finer resolutions, and needs more and more skill to distinguish. When I started in genetics, this was one of the main ways to make a genetic diagnosis: a skilled scientist would look down a microscope and see a subtle change — something missing, something extra, something rearranged. The subtlety of the abnormalities a good cytogeneticist can pick is extraordinary.

When I tried to do this, I struggled even to tell the chromosomes apart, because they don't come all lined up in pairs in the cell — they lie higgledy-piggledy on a glass slide, at all angles, often crossing over one another. It takes at least a year of staring down microscopes at chromosomes under supervision to become a skilled cytogeneticist, and many more years to become an expert. And sometime, perhaps quite soon, technology will make the job, and these skills, obsolete.

It matters a great deal how much chromosome material a person has. Having too much or too little can have serious consequences. Apart from the remnant that is the Y chromosome, 21 is the smallest of the chromosomes, with the fewest genes. Even so, having three copies of chromosome 21 instead of two causes Down syndrome, a complex condition that affects almost every system in the body. Having only a single copy of chromosome 21 instead of two is fatal, incompatible with surviving even the early stages of pregnancy. There are several other syndromes that relate to whole chromosomes. A child with an extra copy of chromosome 18 has Edwards syndrome,[6] for instance. And we've already seen what an extra copy of 13 can do.[7]

The pictures of chromosomes we see have a long history. Early studies of the cell led to the discovery that grasshoppers

---

6   Named for the British geneticist John Hilton Edwards, who described the condition in 1960. This was probably the first example of a genetic condition that was first defined by finding its underlying cause — rather than there being a recognised pattern of features first, with a cause identified later.

7   The X and Y are special in this regard — see chapter 4 for more on this.

have huge germ cells (the cells that become eggs or sperm) and proportionally huge chromosomes, making them easy to study at a time when microscopes were weak and difficult to use. By the beginning of the 20th century, a link had been established between chromosomes and heredity. But it was decades before this promising start was fulfilled and the first firm connection between chromosomes and a human disease was made. For much of the 20th century, we didn't even know how many chromosomes humans have. The number was thought to be 48, not 46, and so everyone who looked counted 48.

Sir Alexander Fleming's discovery of penicillin is the most famous example of lab error gone wonderfully right. Fleming, already a well-known researcher, was studying the bacterium *Staphylococcus aureus*. He went away for a holiday and came back to find that a culture plate that had been left lying around in his (notoriously untidy) lab had been invaded by a mould, and that around the growths of mould were areas cleared of bacteria. Fleming made some progress in understanding the properties of penicillin, including limited attempts to isolate the substance and develop medical applications, but he concluded that this was probably not going to work, and abandoned the line of research. It was left to others — particularly Howard Florey (an Australian) and Ernst Chain (British, but born in Germany), working together in Oxford — to develop the discovery into a usable drug. Despite this, although Florey and Chain shared the Nobel prize with Fleming, they are far less remembered than Fleming.

So you would think that Tao-Chiuh Hsu, who made a similarly serendipitous discovery that had a profound effect on a whole field of medicine, but who then also successfully developed and applied it, would be at least as famous as Fleming. It's one of the vagaries of science history that pioneers are not equally remembered for their contributions — which is why you know who Fleming was, but have never heard of Hsu.

It's a pity, really, because Hsu was a man of remarkable character as well as a great pioneer of science; he should by now have been the subject of at least one biopic. There's a picture of him on the wonderfully named website of the American Cytogenetics Conference, chromophile.org. The photograph was taken in 2000. Hsu is clasping the inaugural ACC Distinguished Cytogeneticist Award, an ambiguously shaped glass object. In this image, Hsu looks like a kindly old uncle. But half a century before that photograph was taken, he was a young man possessed of an adventurous spirit. In the early 1950s, he left China, a very different China from today, to do research on fruit flies (the famous *Drosophila melanogaster*, beloved of geneticists if not of orchardists) at the Texas Drosophila Laboratory in Austin. Texas is famous for many things — the Space Center in Houston, the Cotton Bowl, the Alamo. In a more rational world, the Texas Drosophila Laboratory would be better known than any of them.

Many years later, in a 'mini-autobiography' published in the *American Journal of Medical Genetics*, Hsu praised the openness of the America he had come to, and the willingness of everyone he met along the way to help him. He tells the story of the only incident of racism he ever encountered. En route from the South to a meeting in New Hampshire, Hsu committed an act of heinous driving, reversing on a busy highway after missing a turn. An angry taxi driver paused to deliver his verdict. 'Watch where you are going, you damn Confederate!'

Hsu had come from a technologically undeveloped nation, where cars were rare. He launched himself from a scientific backwater into the forefront of genetic science, a place where he was to stand tall for the decades to come.

The laboratory error in this case happened in 1956, and involved making up solutions that were needed to examine chromosomes. An assistant misread the instructions and mixed one chemical with too much water, making a hypotonic (overly dilute) solution. This caused the cells to swell, allowing the chromosomes to separate so that, instead of being a tangled snarl on the slide, they were much easier to see and distinguish from one another. Hsu seized on this chance opportunity, worked out exactly what had gone so unexpectedly right,[8] and how to consistently achieve the effect, and published his results.

Almost immediately, Joe Hin Tjio (who is remembered, by geneticists at least) and Albert Levan (who mostly is not) were able to use the method to show that in humans, the chromosome number is 46, not 48. If you can't even *count* chromosomes correctly, you can't hope to find chromosomal abnormalities. Now this was possible, and soon afterwards, in 1959, a French team (Lejeune the p persuader, with Marthe Gautier and Raymond Turpin) reported the extra copy of chromosome 21 that is found in the cells of children with Down syndrome. This opened the door to the discovery of many other chromosomal conditions. Even more importantly, better cytogenetics meant the ability to correctly identify individual chromosomes and to create accurate genetic maps. The Human Genome Project, and most of modern genetics, rests on that mistake in Hsu's lab.

As it happened, in the same period, the study of DNA was making real progress at last — Watson and Crick's paper (based on experimental data from Rosalind Franklin) describing the double helix structure of DNA was published in 1953. Their work led to the understanding of the relationship between DNA and proteins described at the start of this chapter.

Between Hsu, Watson, Crick, and the many others on whose work theirs rested, genetics was on its way.

---

8   It took him three months of painstaking effort, changing one step at a time in the process, over the course of countless repetitions, to work out what the difference had been.

2

The DNA Dinner

Yet som there be that by due steps aspire
To lay their just hands on that Golden Key
That ope's the Palace of Eternity

John Milton

I contemplated the two half-full shot glasses sitting in front of me. They were nicely presented on a small wooden board, sitting just beyond my dessert spoon. Everyone at the table had the same arrangement. My pre-dinner drink was still half full, and I didn't particularly want to go straight to shots — but, on the other hand, there was a mood of celebration in the air, and it was a little tempting.

It's just as well that neither I, nor anybody else in the large, crowded hotel ballroom, succumbed to that temptation. That glittering occasion — the 50th anniversary of the discovery of the double helix structure of DNA — may not have been quite the same had a dinner guest been poisoned.

The DNA Dinner was a highlight of the 19th International Congress of Genetics, held in Melbourne, Australia in 2003. Its organisers were people who were open to ideas, and not afraid of consequences. This made for a night that was thoroughly memorable, if a little hazardous. Huge helices of balloons coiled up the walls, as you might expect. Not so predictably, whoever was responsible for them had apparently decided that the double helix looked a bit thin, so the balloon decorations paid tribute to the triple helix instead.[1] Francis Collins, leader of the Human Genome Project, played the guitar and sang 'Happy Birthday to You' ... to the human genome.

And then there was the poison.

To be fair, apart from the potential for harm, it was a brilliant idea. One of the glasses contained a nearly complete extraction of DNA from a plant. The other contained the last chemical needed to complete the extraction. At the beginning of the night (probably just in the nick of time), the MC told us to pour the contents of one glass into the other — and before our eyes, DNA precipitated out. There it was, in front of us — the magical stuff that had brought us all together.

You can do your own DNA extraction at home, using household ingredients (only one of which is poisonous). I'm told that cookbooks sell well — so here's a recipe.

Ingredients
- Several strawberries (two or three will do, depending on size)
- Water
- Salt
- Dishwashing liquid
- Rubbing alcohol (isopropyl alcohol); methylated spirits can be substituted

---

1   The triple helix is important, too — collagen, one of the major types of protein that make up your body, has a triple helix structure. But this wasn't the Collagen Dinner.

Preparation

1. Add one teaspoon of salt to half a cup of warm water, and stir until dissolved.
2. Add two teaspoons of dishwashing liquid to the salty water. Stir gently (you don't want it to froth).
3. Put the strawberries in a ziplock sandwich bag and seal. Mash the strawberries inside the bag thoroughly by hand.
4. Pour the salty, soapy water into the bag with the strawberries.
5. Mix well. Be gentle, avoiding frothing of the mixture.
6. Use a coffee filter to strain the mixture from the bag into a glass. Make sure you get a decent amount of liquid in the glass. It's best not to use a tall, narrow glass for this step because it will make life difficult at the end.
7. Pour rubbing alcohol gently down the side of the glass, in about a 1:1 ratio with the liquid from the strawberry. The alcohol will form a layer on the top.
8. Now it's poisonous. Do not drink.
9. Allow to stand, and wait. DNA will rise into the alcohol layer, in a gloopy white mass.

At this point, if you like, you can take a wooden skewer and lift the DNA out of the glass. It has some interesting properties. Dab it against the side of the glass a few times — the clump on the end of your skewer will get smaller. Lift it carefully enough out of the liquid and you may be able to get a long, thin strand to rise up from the surface. DNA is sticky, and can be loosely or tightly coiled. When it shrinks, that's loose DNA coiling more tightly as it sticks to itself. The string of DNA you can pull up from the surface of the liquid is a series of individual strands, sticking to each other as they are lifted out of the container.

I *strongly* urge you to give this a go. It is deeply satisfying to know that you are holding the stuff of life in your hands. Even if it does look and feel rather like snot.[2]

---

2   No, I don't know what it tastes like.

Francis Collins was the focus of attention that night at the DNA Dinner, not because of his guitar playing and singing (fine though they were), but because of his leadership of the Human Genome Project. Three years previously, on 26 June 2000, there had been a ceremony at the White House, announcing that the human genome had been sequenced. President Bill Clinton hosted the event, and British prime minister Tony Blair joined by satellite. Collins and the publicly funded Human Genome Project shared the limelight that day with a private company, Celera. A remarkable effort, led by Celera's CEO, Craig Venter, had turned the sequencing of the human genome into a race, ending in an effective dead heat between the public and private efforts.

Some might say that the ceremony jumped the gun a bit, because there were still so many gaps in the sequence — no fewer than 150,000 of them — and because at least 10 per cent of the sequence was still missing. In fact, there was another announcement on 14 April 2003 that the project was *really* finished, but, even then, there were still gaps. By 2004, things had improved considerably, yet there were still 341 gaps ... and even today, the job is not quite finished.

Nonetheless, at the time of the announcement in 2000, there was a good working draft — and to be fair, that was what was announced: completion of a draft sequence. Most researchers, most of the time, could consult the data with the expectation that there would be detailed information available about the region they were interested in. It was an exciting time, but it still wasn't entirely clear to those of us in the clinical world exactly what we were going to use the genome sequence *for*.

One day in late 2001, a package arrived in our department that was to thoroughly prove this point. It was the human genome on disk — sent to us, free of charge, by Celera. Excitedly, we opened it up, loaded it onto a computer, and started exploring the contents. And quickly stopped. We had no idea how to interpret the information we had been sent, and no way to connect it with our patients. As it turned out, it would be more than a decade before genomic data would become a routine part of practice in clinical and diagnostic-laboratory genetics. Nowadays, I consult the genome browser run by the University of California, Santa Cruz numerous times during the course of a working week. I could not do my job without it.

So — what's in the genome? What exactly am I browsing, thanks to UCSC?[3]

That white gloopy stuff you extracted from the strawberries is made up of four different chemical building blocks: adenine, cytosine, guanine, and thymine. Their initials are A, C, G, and T, and they are called nucleobases, or bases. The human genome is composed of about three billion bases altogether. Most of the time, they exist as base pairs, because of the double helix in which DNA usually exists. That double helix is made of two individual strands, which complement each other. A on one strand sticks to T on the other, and C on one strand sticks to G on the other, so that the double helix looks like this:

```
G A T T A C A
| | | | | | |
C T A A T G T
```

The two strands run in opposite directions — there is a direction to DNA, related to the way it is copied and translated to make proteins. So the strand that complements 'GATTACA' would be read as 'TGTAATC' by the cell's machinery, not 'CTAATGT'.

Three billion bases of DNA is an awful lot. To put that in perspective, here's a chunk of human genetic code:

```
GAGGGGTGACAATAGGAATATTTGCTTTTCATCCCTCATGATCAT
CACCCTTCCCTTCTTCCACTCTTACATTTTTATTTCCCAAAGT
```

---

3   I'm afraid you might find my browser history a little dull.

GGGCTGCAGTACTGGTGGTGAAAAGCAGTCATTCTAGAACT
CACTAGCTTGGTCAGGAAACACTGGGCATGCTACTCCGTGGGAATC
CCAGGTAATTTTATACAGTGATGATGGGTCATCCCTACAGCCTG
CCTGGAGTTTCTGAGCTTAACTTTTCCTGAGTCAAATCCAAGTGAC
CATTTGGTTATGCTGTTCTTTCCAGATCTTTCCTTGTTGAACAGT
GTCAGCTGTCATTCACAATTCTCTTTACCTCCAG**AGCAATTTGT**
GGAGAAGTCGTCCTGTGCCCAGCCCCTGGGTGAGCTGACCAG
CCTGGATGCTCATGGGGAGTTTGGTGGAGGCAGTGGCAGCAGC
CCGTCCTCCTCCTCTCTGTGCACTGAGCCACTGATCCCCACCACCCCCA
TCATCCCCAGTGAGGAAATGGCCAAAATTGCCTGCAGCCTGGAGA
CCAAGGAGCTTTGGGACAAATTCCATGAGCTGGGCACCGAGATGATCAT
CACCAAGTCGGGCAGGTAGTTGGGACTTGGGGGTTGGGGGTTGGGG
GTGGAGAGTCAGGACACTCCCTGGGTAGTTGAGGGTGCTTCCAG
GAACTAGATGAGAGCTGGCTGGTCATGGAGCGGAGAGACAGCTTG
GCTCCAGGGCAGCTGCTTTCCACCAGCTTGCATTAGGAGCTACAG
GATGTCTAGTCATTTGCGTTCTCAGGATTTGGTCATGGGAAG
CCCCACCCTGGCTTTGTTGAGAAGGGCACAGGGACCAGGGAG
ACACACTAACCCCGAAGGGTGTGGTCTGCTTTCCCTGGAGCTG
GAGAAGGTTTGGCGGGTGGAGGGTCGGGATCTGGAAGGAGGAG
GAATTTGTGCCTGGGTGCCTGGTGAGCTGCTGGGTGCTTCTAGG
TAGGTGAGTAGCTTCCCTTTTATCAGCCTCAATTTGCAAAAGCTG
CCAGCTCCCATTAAAAACTAAAATTAAAACCTGGGCGGAAGAAT
GAAATTTGAAACGATAAAATTCCCTGTAGGAAGGAGCACTGCTC
GGGGCCTCTTGGCGCCAGAGCCGGGCGGGCTTTGGCCAGGCAG
GAAGCTGCAGGGCTGCAGGGAGGTTGGGATGGGGCAGAGGCTGG
CAAAACTTGGTGGCTCTAGCTCTTGGGACTACAGAAAATACCT
GCAGGGCATCTGAGAAATCCTTCCCAGAAACCTCTGCTTTTG
GCTTTTATTTTGCAAGAGCAGAGTTTTCTGGCTGGGATGCGGGT
GAGTTGTGTGACTGGGTCAGCTCCAGGGACTTCGGGTCCTGGGA
CACTTAATGTGCTTGATCGTTAAAATGCATGGGATTTTCCCTA
ATCACAGACCTTCTGGAGTTAACACATACCCCCACCCCCAC
CCCCACCTTTTCACCTAGCAATTAACACCTGCTTAAAGGTGA
CACTTAAAATTATCTAGGCTTGGAAGAAAACCCTGTCTCTGTAT
TCACTTCTCTGAGGCTTTAAACAAAACAAAAGAGGGGTTTGTG
GACCGGATAGAGAGGGGAGTCAGACCCTTTCCCTCCTTCCCTC
CCTCCCTTCCTCTCTTCCTAATTCAGGTCAGTTTATTAGGCAG

```
CATAAACAGGGCCCATTCTCTCTCTCTCTCTGTCAGGAGGAT
GTTTCCAACCATCCGGGTGTCCTTTTCGGGGGTGGATCCTGAGGCCAAG
TACATAGTCCTGATGGACATCGTCCCTGTGGACAACAAGAGGTACCGC
TACGCCTACCACCGGTCCTCCTGGCTGGTGGCTGGCAAGGCCGACCCGC
CGTTGCCAGCCAGGTTCGTGCCTCCAGATTTTTCACTGAGAAAACT
GTTAGTGCATCTGTCAGAATGTTTCTGGCTTGTGTGAATTTTAAG
CAAGTGTATTTTTAAAGCAGCGGGCTCTGGCAAGAGAGCATTC
CAAGCCTGGACACTCCAGGATTGACTACACAAAACATGGGCTAG
GCTCTGAGAAAGGTAGTTTGTGCATAGAGAAAACACTGTCTTTA
AGTTTATGTTTCGTTAGGCAGTAATTCATTTCAAAGTTTTTCTTA
AATTTCAATTTGAGTATTCATTAGAAATGTGGACCCATTTTGTATA
AATATAAATATAGACATCCTCTCTAATTGCTGCTTAAAACCAGAGTGAA
```

This is one of my favourite bits of the genome — it's part of the gene *TBX20*, which had a starring role in my PhD. Printed at the same density on A4 paper (single sided), you'd need 781,250 sheets of paper for the whole of the human genome. If each sheet of paper is 0.1 mm thick, you would need a stack of paper just over 78 metres tall (198 feet) — halfway in height between the Sydney Opera House and the Statue of Liberty. Without a key, of course, this would be just a stack of meaningless letters. *With* the key — that stack of paper contains untold scientific wealth.

What is the key? And what's *in* the genome? It turns out that it's more a matter of a *set* of keys, rather than just one. DNA tells many stories, if you can read them.

As we saw in the previous chapter, we have pairs of chromosomes[4] — pairs, because you get half of your genetic information from your mum, and half from your dad. In turn, you pass on half to each child you have. So — one copy of chromosome 1 from mum, one from dad, and so on. Chromosome 1 is the largest. It's about a quarter of a *billion* bases long, and is home to more than 2,000 genes. The smallest, chromosome 21, is less than 50 million bases long and holds only a couple of hundred

---

4   Feel free to go back and admire my chromosomes again at this point.

genes. The humble Y chromosome is just a little longer than chromosome 21, but has only about 50 genes.

There is also some DNA outside the nucleus of the cell: we have a second genome, a tiny one (only 16,569 bases and 37 genes). It lives in structures called mitochondria — more about them later.

Speaking of genes — you've doubtless heard of them, because they are the most famous things in the genome. As I explained, their job is to act as a blueprint that tells the cells how to make proteins, which in turn do the many complex tasks a cell needs to do in order to be alive and contribute something useful to your body. The parts of the genes that get translated to make proteins only account for around 1–2 per cent of the genome.

There's still quite a bit of controversy about how much of the rest of the genome actually does anything. Some of the non-gene bits are definitely useful and important. For example, the centromere, at the waist of the chromosome, is essential for making sure the chromosome copies go where they are supposed to when cells divide. Mess that up and the consequences are not good. The ends of the chromosomes have structures called telomeres, which form a protective cap. You may be familiar with the Bernard Bresslaw song about feet:

> You need feet to keep your socks on
> And stop your legs from fraying at the ends

Chromosomes are not known for wearing socks — but like your legs, it's bad news if they fray at the ends. As you get older, the telomeres themselves do tend to fray a bit, gradually getting shorter with each cell division. During the development of many cancers, they wind up a *lot* shorter than they should be, or disappear altogether, leaving the ends of the chromosomes exposed and vulnerable to damage. Paradoxically, what happens next is a restoration of the telomeres — as cells are undergoing transformation to malignancy, their chromosomes reacquire robust telomeres. This is part of what makes a cancer cell 'immortal'.

Although the parts of genes that code for proteins only account for about 1–2 per cent of the genome, genes spread across about a quarter of the genome. The reason for that difference is that most genes are a mix of two types of sequence — introns and exons. The exons code for protein, i.e. their sequence specifies which amino acids to include in the protein, as well as when to start and stop. By contrast, the introns don't code for anything, and, while they undoubtedly still do serve a purpose, we still only have a fairly limited understanding of what that is.[5] Introns can be truly enormous — many thousands of bases long. Sometimes, they are so big that there's room for an entire gene to sit inside an intron of another gene, usually running in the opposite direction, on the other strand of DNA. The double helix is a two-way street.

You can see exons and introns in that chunk of *TBX20*. The bold sections are exons; the bits in between are introns. You can even see some of the genome's operating instructions, written right there in the DNA sequence. At the start of each intron are the bases GT; at the end of each intron are the bases AG. Together, GT and AG form a key part of a message to the cell's machinery that says, 'There's an intron here. Not needed for protein — please cut out.'[6]

What percentage of the human genome actually does

---

5   One well-understood function of introns is to enable the use of the same gene to make different versions of its protein, sometimes versions that have quite varied functions. This is done by 'alternate splicing' — some exons don't always get used — so there is sequence that could be either exon or intron. Plenty of genes don't do this at all, but there are some proteins that come in numerous flavours depending on the way that splicing happens. Another function of introns is to help control when and where a gene is switched on — a regulatory function.

6   This is something of a simplification. The GT and AG are key parts of the signal that says, 'I'm a splice site,' but the bases that surround them are important, too. There's a more detailed description of the relationship between gene and protein in the notes section, if you'd like to read more about this.

something? Well, we *still* don't know. In September 2012, the results of a major follow-up to the Human Genome Project, called the ENCODE project, were released in 30 scientific papers, all in one hit. Getting that many scientists to cooperate so that 30 papers were released into the wild simultaneously was perhaps as impressive an achievement as the actual science in the papers. The ENCODE Consortium reckoned they had found a function for 80 per cent of the genome. Most of it was claimed to be busy controlling the function of other bits — a rather bureaucratic vision of cell biology. There was a lot of criticism of this announcement at the time, and the debate rolls on. Recently, a paper was published that argued that only 8 per cent of the genome is functional. That's quite a gap. I have no idea what the right answer is, but I doubt it's as low as 8 per cent or as high as 80 per cent.

There's an awful lot of the genome which looks like genetic wreckage — genes and other elements that have lost their function over the course of evolution. For instance, there are loads of smell receptor genes that are broken and don't do anything — earlier in evolution, our ancestors needed a keen sense of smell to survive, but for a long time we've been able to get by just fine with a comparatively poor sense of smell. So when those genes acquired mutations, it didn't cause a problem, and the broken form was just passed on to future generations. You inherited hundreds of broken genes from your parents, and in turn you'll pass them on — or have already — faithfully copied, and still not doing anything.

There are also lots of repetitive sequences that don't seem to be up to much. Sometimes, viruses copy themselves into a host's DNA, and there's quite a lot of what look like old virus sequences scattered about the place. There are chunks of DNA that have been copied in what's called duplication events. If you have spare copies of something, it doesn't matter much if one of them loses function, so you often wind up with two versions of a gene, one that works and one that doesn't, a pseudogene. And there are bits of DNA that seem to be there simply as a side effect of DNA's drive to be copied: long, long strings of sequence that looks like nothing much (ATATATATATATATAT …).

On the whole, this doesn't seem to cause us any bother. There's no apparent pressure on the human genome to get more efficient, or maybe whatever pressure exists is outweighed by DNA's tendency to copy itself, and by the various mechanisms which introduce new bits of DNA into the sequence. There are plenty of other organisms that get by fine with much bigger genomes than ours, with proportionally even more freeloading DNA. There's an amoeba, *Polychaos dubium*, that reportedly has a genome more than 200 times the size of ours. The humble onion's genome is five times bigger than ours, and, on the whole, you are more likely to eat an onion (or extract its DNA) than the other way around. On the other hand, the pufferfish, fugu, has a genome only about an eighth the size of ours … and pufferfish are quite a bit more complex than onions.

It does seem like there might be a price to pay for an over-sized genome, at least when times are hard. There's a plant called teosinte, which is thought to be the ancestor of maize. In 2017, a paper was published that compared the size of the genomes of different species of teosinte, living at different altitudes. Many plants have enormous genomes — but in teosinte, at least, the higher the altitude, the smaller the genome. If you're living in a tough neighbourhood, high on a mountain, you can't afford to waste energy copying DNA if it isn't doing a valuable job for you.

It's just about possible that the human genome happens to be exactly the perfect size, so that everything in it has an important role. But that would seem like quite a fluke. More likely, the human genome does indeed carry around its fair share of true 'junk' DNA.

That's not to say there isn't anything impressive and interesting about the human genome. When I started working in genetics, we used to confidently tell people that the human genome contained about 100,000 genes — because we're such important, special creatures, there had to be a lot, right? Then that estimate started going down ... and down ... and down. By the time the Human Genome Project was completed, the number had gone down to a bit over 20,000 genes. Part of the explanation for this is that our genes have quite complex structures, and an awful lot of them do more than one job. Sometimes, that means doing a similar task in slightly different ways, like a muscle protein that forms a bit differently depending on whether it is working in heart muscle or in regular muscle. Sometimes, though, it means the same protein can be used to do wildly different jobs. This is called 'moonlighting'. For instance, there's an enzyme — a protein that makes chemical reactions happen — that also plays an important role in making the lens of the eye transparent.

But you can say similar things about the genomes of a lot of organisms, and *all* of it about the chimp genome. Chimpanzees, especially the bonobo (or pygmy chimpanzee), are so close to us genetically that a Martian would probably view us as just different varieties of the same animal. We are closer to chimps than African elephants are to Asian elephants, so you could hardly blame our extraterrestrial visitor for being confused.

How do we know all this stuff? It comes back to the Human Genome Project.

At the time of its conception, the HGP was an enormously ambitious idea. Only a small fraction of the genome had been sequenced. Mostly, what we had was a kind of outline sketch — a map, in fact. You often hear the expression 'mapping the genome', and indeed that was the first step. But we never map a person's genome now, because the job has been done already — just as you don't have to map someone's whole neighbourhood in order to find their house. A genetic map is not like a street map, because it only really has one dimension, not two — it's all about what lies where along the string of DNA that makes up a chromosome. To make such a map, you need a series of markers — genetic signposts — with a known relationship to each other. Those signposts consist of bits of DNA with some way of uniquely identifying them. Say we have three such markers, A,

B, and C. If we make a genetic map that includes A, B, and C, it would consist — at the least — of the information that they sit along the chromosome in that order — A-B-C — rather than A-C-B or any of the other possibilities. A better version would say that A, B, and C are all on chromosome 1, and not on any of the other chromosomes. And the most useful type of map would also tell us how far apart they are.

The earliest such maps were made in the early 20th century, for fruit flies. By 1922, genes for 50 different characteristics had been mapped to the four fly chromosomes. These were all physical differences in the fly that a researcher could directly observe. Flies would be examined for multiple different characteristics and mated with flies that had also been carefully examined, and the resulting offspring would be examined in turn. It was exacting, difficult work, but taught us a lot of fundamental information about genetics, and gave us tools that were used all the way through the 20th century, and were essential to the success of the Human Genome Project.

For the fruit fly's X chromosome, for example, there was an early map that looked like this:

In this map, y stood for yellow body, w for white eyes, v for vermilion eyes, and m for miniature wings. The map means that yellow body and white eyes are closely linked — they are more likely to be inherited together — whereas the miniature wings variant is more likely to be inherited along with vermilion eyes than with white eyes. This particular map was the work of Alfred Sturtevant, another nearly forgotten genius, and was put together in 1913, when Sturtevant was only 21. At the time, he was working under the supervision of the great geneticist Thomas Hunt Morgan. Sturtevant seems to have been a child prodigy: by age 21, he already had a long track record of studying inheritance. Morgan had been impressed by Sturtevant when, still in his teens, he wrote a paper about how horses inherit coat colours — based on observations made as a child, on his father's farm! The paper was published in a scientific journal, Morgan offered Sturtevant a position in his lab — and the rest is history.

That's some school project.

Sturtevant went on to have a long and distinguished career in science, pausing only to marry Phoebe Curtis Reed, a technician who also worked in the fly lab. They had three children, who must have grown up with quite unusual ideas about what constitutes a normal topic for dinner table conversation.

From the point of view of genetic mappers, most of the 20th century was a hard slog. From mapping physical characteristics that you could see, things moved on to biochemical and other laboratory markers — in yeast, and eventually in humans. It wasn't until 1987, 17 years after Sturtevant died, that the first genetic map spanning the whole human genome was published — 407 variable bits of DNA spread across the 23 chromosomes. If the Human Genome Project was the moon landing of genetics, Sturtevant's early maps were our Wright brothers' flight.

So, by the late 1980s, we had that outline map, like those early explorers' maps of the world. We knew the outlines of our 23 land masses, and there were definite signposts spread along them (those 407 genetic markers), but, apart from a few important ports — chunks of known DNA sequence centred around disease genes — there was precious little detail to be found on our maps.

Going from a 407-marker sketch to a richly detailed map, and then to a completed genome sequence, needed some special tools. One of the most important of those was Sanger sequencing.

You've surely heard of Marie Skłodowska Curie, who won two Nobel prizes, in physics and chemistry. There's a good chance you've heard of Linus Pauling (Nobels for chemistry and peace). I have to confess I had never heard of John Bardeen until I started writing this chapter. This is embarrassing, since it seems we all owe him an enormous debt for the work that won him two physics Nobels. Bardeen was co-inventor of the transistor, and developer of the theory of superconductivity — your phone only works because of Bardeen's discoveries. The same goes for the computer on which I am typing this.

But for our purposes, there can be no doubt that Fred Sanger was the greatest of the quartet of double Nobel winners. Sanger, an Englishman, was a chemist who worked out how to determine the sequence of amino acids that make up a protein, for which he received his first Nobel. Sanger was a Quaker who received official conscientious-objector status during World War II, and a good thing, too — it would have been an appalling loss to the world if he had died during the war.

The protein Sanger chose to study first was insulin — the sugar-regulating hormone that is lacking (or ineffective) in people with diabetes. Insulin had been successfully used to treat diabetes since the early 1920s, and was one of very few proteins available in pure form in the early 1950s. The story of how that came to be is a special part of science history in itself.

Media stories of medical breakthroughs tend to be evenly divided between breathless reports about small improvements that happened several years ago and research in animals that might never translate into something relevant to humans. My PhD supervisor, Richard Harvey, and I were once interviewed on national TV news about research *we hadn't even done yet*. We'd been awarded a grant from the National Institutes of Health in the US, and our institute's public-affairs department had somehow sold this as a big news story to one of the TV networks. Years later, when we had done the research and had the results published, we couldn't even get a local newspaper to give us a mention.

But there are some genuine stories of miracle cures in the history of medicine. The introduction of penicillin is one: deadly, incurable infectious diseases suddenly became curable. But for a real wonder drug, nothing beats the story of insulin.[7]

---

7   Okay, there is *one* that's better. Anaesthesia beats insulin. And I am definitely not saying that because my wife is an anaesthetist.

Diabetes comes in two main flavours — and I use the term advisedly. Diabetes mellitus, by far the most common type, gets its name from a Latin word meaning 'sweetened with honey' — because the urine of sufferers tastes sweet. If you were to taste the urine of someone with diabetes insipidus, by comparison, you'd find it insipid: flavourless. The urine sommelier at your favourite restaurant would certainly not recommend it.

In turn, diabetes mellitus is divided into two broad groups, based on how well treatment with insulin works. If you have a deficiency of insulin, because your pancreas has stopped making it, then what you need is a replacement, which you can get in the form of an injection. This is the type we'll be focusing on here. On the other hand, if your body makes insulin fine, but no longer responds normally to its effects, you have non-insulin-dependent diabetes, quite a different problem. As you might expect, there are all sorts of subtypes beyond this broad division. One rare type that affects newborn babies will make a cameo appearance later in the book.

The main job of insulin in the body is to give your cells the go-ahead to take up and use glucose from the bloodstream. If there's no insulin around, it's as if your cells are sugar-blind — they just can't tell that the glucose is there, and they can't do anything with it. Since the sugar isn't being used, it builds up in the blood and spills into the urine, making it sweet — and dragging water with it so that you pee too much and get dehydrated. Meantime, your body is starving in the midst of plenty, because your cells can't use the glucose that's there.

By the early 20th century, it was already known that if you removed a dog's pancreas, the animal would develop diabetes and would die within a fortnight. Unfortunately, much the same was true for people — diabetes, which mainly came on during childhood, was a death sentence. Affected people might linger on for a few weeks or months, but eventually they would slip into a coma and die.

There are several heroes in this story, all Canadians. Frederick Banting was a surgeon who had an idea about how to get an extract from the pancreas of a dog that might be used to treat diabetes. People had tried to do this previously, but nobody could get it to work.

As well as making insulin, the pancreas makes digestive enzymes, and Banting thought that, when people mashed up pancreatic tissue to try to extract insulin, those enzymes were coming into contact with the insulin in the mash and digesting it, so that there was none left to be extracted. His idea was to tie off the ducts that carry the digestive juices from the pancreas to the gut, causing the cells in the pancreas that made those enzymes to wither away. He hoped that, when this happened, you could mash up the remaining tissue, which would mainly be the insulin-producing cells without the enzymes, making it possible to get a pure preparation of the stuff they were after. He went to a leading diabetes researcher at the University of Toronto, John Macleod, who took some persuading but eventually gave him the resources he needed to give the idea a go, including ten dogs and an assistant, a medical student called Charles Best. Interestingly, it's still true that, in the medical hierarchy, medical students rank slightly above domestic animals.

The story goes that Best was the winner of the luckiest coin toss in medical history. There were originally two students who could have been assigned to the project. Best and the other student, his friend Clark Noble, tossed a coin to see who would work with Banting, and Best won. Initially, they were going to swap halfway through the summer, but by then Best was entrenched in the work (and technically skilled at doing it), and they agreed that he should stay on. A share of scientific glory, on the toss of a coin.

It turned out that Banting's idea was right. The initial work in dogs was so encouraging that, by January 1922, it was possible to start trials in humans. Another figure enters the story here: James Bertram Collip was the one who worked out how to purify the pancreas extract so that it could be safely injected into people. He was called in after the first recipient suffered from a severe allergic reaction when given the imperfectly purified extract of dog pancreas. There are stories from this time of extraordinarily dramatic recoveries. In particular, it's said that once they had a pure preparation, Banting's team went from bed to bed in a room full of comatose, dying children, giving injections. By the time they reached the last one, the first had already awoken from his coma.

If this story is true, you certainly can't tell from the first scientific report of treatment with insulin, published in the *Canadian Medical Association Journal*. This paper is breathtakingly, gloriously dull. It's not until halfway through the second page (in a paper only a little over five pages long) that there is even a *mention* of treatment in humans. The conclusions are cautious and measured: in essence, 'we can measure some differences in the blood of patients, and they seem to feel better'.

Even if Banting and co. were reluctant to blow their own trumpets, word of such a discovery was bound to get out, and the news raced round the world. The very next year, Banting and Macleod were awarded the Nobel prize for medicine or physiology. Banting shared his prize money with Best; Macleod shared his with Collip. We can only imagine what it must have been like for families of newly diagnosed diabetics, after the news was out, but before large-scale insulin production was possible. There must have been many people who died despite the existence of a treatment, and others who were saved in the nick of time.

Be that as it may, by the time Fred Sanger needed a purified protein to study, 30 years later, all he had to do was stroll down to the local pharmacy and buy a bottle. Sanger's approach was to simplify the problem — instead of trying to read the sequence of the entire protein, he would smash it into shorter bits that were easier to handle. He developed chemical methods to work out the sequence of amino acids in those short sections, then pieced together those overlapping short sequences to work out the overall sequence of the protein. It was an idea that had a long scientific reach. As we shall see, Craig Venter's company, Celera, would use essentially the same approach to sequence the human genome, nearly 50 years later, and it remains an important technique in genetics. In 2018, the koala genome was sequenced this way.

Sanger's discovery was not just 'this is the sequence of insulin'. Important though that would have been, its impact was trivial compared with the discovery that proteins *have* a set sequence, on which their structure and function depends. Many later discoveries, including our understanding of the way that DNA codes for proteins, would have been impossible without this basic understanding of what a protein actually *is* — a chain of amino acid molecules.

For his next trick (and next Nobel Prize), Sanger figured out a way to read the sequence of DNA. His method, published in 1977, is still known as Sanger sequencing, and was the foundation stone on which the Human Genome Project rested. At first, Sanger sequencing involved a certain amount of messing around with radioactive isotopes, but later improvements involved tagging the DNA bases with different coloured fluorescent markers — safer, and also possible to massively scale up. And scaled up it was.

We still use Sanger sequencing in diagnostic labs. Here's an example of what it looks like:

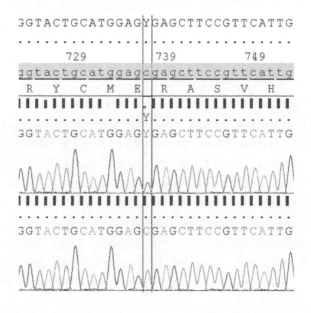

In this picture, the top section is from a carrier of a genetic condition and the bottom section is from someone who isn't a carrier. You can't tell from a black and white image, but the convention is that A is green, C is blue, G is black, and T is red. Each of the different coloured peaks represents one base of DNA, so you can use those colours to read the sequence from the peaks … G, G, T, A, C, T, and so on. Or you could just read the sequence of letters helpfully placed above the peaks, I suppose — but you see how it works.

In the middle of this stretch of DNA, between the two vertical lines, the normal sequence has a C; at the same place in the sequence above, if the image were in colour, you would be able to see that there's a red peak there, a T — but if you looked very closely, you would see that the normal blue peak, the C, is there as well. This means that one of the two copies of the gene has the usual sequence, and the other has a change. Both copies, the one with and the one without the change, are in the sample and undergo the sequencing reaction. For the bases that aren't changed, there's no difference between the two, and the resulting trace looks the same as the normal sample. For the altered base, half of the DNA in the tube has a C and half has a T, and the two overlap — that's why the peak at that place is about half the height of the C in the normal sample.

That's easy enough to do now because we already know the sequence of the gene that we're interested in. But it's also possible to use Sanger sequencing to discover a DNA sequence where it wasn't known before. You start from a known section and work your way along, discovering new territory as you go — until you meet up with someone coming the other way. This was the approach used by the Human Genome Project.

So, by the late 1980s, we had the tools that would be used to complete the job. Actually completing it, however, seemed a very long way away. In 1987, the US Department of Energy, an organisation not known to be daunted by large-scale ventures, launched an early version of the project. Their agenda was to find a way to protect the genome from the harmful effects of radiation — potentially important information at a time when nuclear power plants were an important and growing part of the country's energy supply. By 1988, the National Institutes of Health had joined the DOE, and together they had received funding from the US Congress to attempt the task.

It would be a scant 12 years before this early vision would become reality. Yet for the first half of that time, from the perspective of an outside observer ... almost nothing happened. In 1990, a goal of 2005 was set for completion of the project, but, given the expectation of 'over time, over budget' for government projects, not many outsiders took this very seriously. By 1994, the main achievement of the HGP was ... a denser map. The new map of the genome had not 407 but 5,840 markers, densely spaced across the genome. To the outside observer, not that impressive, perhaps, but it was a critical step on the road to the genome. And in a sign of what was to come, it was delivered a year earlier than expected.

From very early on, the HGP was an international effort, with scientists from all over the world contributing. An Australian, Grant Sutherland, was president of the Human Genome Organisation (although not leader of the Human Genome Project) for part of the project's life. James Watson (yes, that Watson) was the first leader of the HGP. In 1992, he resigned, and, after a brief interim, Francis Collins took charge, and would see the project all the way through to its end.

The actual sequencing was done by 20 institutes spread across six countries — the US, the United Kingdom, Japan, France, Germany, and China. Each was assigned a section to work on. The biggest single contributor outside the US was the Sanger Centre at Cambridge, named of course for Fred Sanger. The Sanger Centre, now the Wellcome Sanger Institute, sequenced almost a third of the human genome. Their assigned turf — some shared with other institutes — included chromosomes 1, 6, 9, 10, 11, 13, 20, 22, and X. The project seems to have gathered speed like the proverbial runaway locomotive. It wasn't until 1999 that the first complete sequence of a human chromosome (chromosome 22) was published. In September 1999, more than a decade after the project started, a press release trumpeted the completion of 821 million bases of DNA sequence. Half of that was still in 'draft' form, and more than two billion bases still needed to be sequenced before the project reached its goal. But by June the next year, the project was near enough done for President Clinton and Prime Minister Blair to claim success.

The HGP took a logical, safe approach to sequencing the genome, taking what was known as 'walks' from known to unknown places. But they had a competitor — J. Craig Venter, the Elon Musk of genomics, who proposed a different approach altogether. Venter wanted to use a shotgun to get the job done.

Venter was a brilliant scientist with an entrepreneurial streak. After a less than stellar school career, from which he emerged a better surfer than scholar, Venter was drafted, and he served in the United States Navy during the Vietnam war. Experiences in a field hospital had a profound effect on him, and, when he returned, he studied medicine, but later switched to research. He proved an outstanding scientist, but, while working for the National Institutes of Health, became involved in a controversy over efforts to patent genes, and later left the NIH to go to the private sector — where he also excelled. As the first president of Celera Corporation, he decided to race the Human Genome Project, using a method the HGP had rejected — shotgun sequencing. This involves smashing the genome up into many small pieces, sequencing those, and then assembling them like a giant jigsaw puzzle.

Let's say you do some sequencing and find that you have three fragments like this:

G G T G T G A A C T G C C C C G A G G G
C C G A G G G C A G A G A C C T C C G T T T T G
C G T T T T G T T C T C C A G C G C C T T G A G C C A G C

With the right computing oomph behind you, it's possible to put these together, like this:

G G T G T G A A C T G C C <u>C C G A G G G</u> C A G A G A C C T C
<u>C C G T T T T G T T</u> C T C C A G C G C C T T G A G C C A G C[8]

The first chunk overlaps with the second, and the second with the third. Without the second, you'd have no way of connecting the first with the third. But keep on smashing and sequencing, smashing and sequencing, and eventually you'll have enough overlapping bits that you can put the whole thing together. And in a genuinely astonishing achievement, Celera did exactly that. Pitted against 20 institutes in six countries *and* the US Department of Energy, Celera crossed the finish line in lockstep with the public HGP, which is why Venter shared the floor with Collins that day at the White House.

Celera did have one important advantage over the HGP — access to all of the public body's data. From the beginning, one of the fundamental principles of the HGP was open access to data, setting the tone for a standard that continues in bio-medical science to this day.

You may wonder what Celera hoped to gain by this effort. The initial plan was to discover and patent genetic sequences. Celera did file preliminary patents on 6,500 gene sequences, but in the end did not follow through with the patent process, and made their data freely available as well (including sending us the disk that, through no fault of theirs, we couldn't interpret).

Even from the beginning, the 'reference' human genome was

---

8   Another non-randomly chosen sequence: this is from *NKX2-5*, which also featured in my PhD and is one of my favourites for another reason. Animal geneticists in general, and fly geneticists in particular, have always been much better than human geneticists at naming genes. *NKX2-5* is important in the development of the human heart. Flies don't have much of a heart — just a tube that squeezes, really — but they have a gene that is very similar to *NKX2-5*. When that gene was found, it was discovered that flies that lack the gene don't develop their tube-heart at all. The name of the fly gene? *Tinman.*

a mishmash of the genomes of different people — as it should be. The HGP called for volunteers from people living near the 20 sequencing centres. No personally identifying information was kept, and only a small proportion of the samples collected were actually used for sequencing, so nobody knows whose DNA is the 'reference'. The 'reference' genome we work with today has been updated and adjusted using extra information from many different people; it's a quilt made from many patches of cloth of varying sizes. Celera did something along the same lines, although perhaps less random. Twenty-one donors were enrolled. Apart from age, sex, and ethnic background (self-described), no information was kept about them. The volunteers had to provide 130 mL of blood (a little over a quarter of a pint, enough to make for quite a vivid crime scene). Males also provided five specimens of semen, collected over a six-week period (the section of Celera's *Science* paper that describes their methods is strangely interesting). Of the 21 volunteers, just five people were chosen for sequencing. Two men and three women; one African American, one Chinese, one Hispanic Mexican, and two Caucasians.

It turned out later that one of the Caucasians, a male, contributed disproportionately to the effort, and was no longer anonymous: it was Venter himself. Only a few years later, Venter had the remainder of his genome sequenced, possibly the first individual human being to have this done. 'Possibly' because, at around the same time, James Watson had his genome sequenced, and it's not clear which was finished first.

That was in 2007. At the time, sequencing an individual human's genome was an astonishing idea. Now, it's almost commonplace — you can have your own genome sequenced, if you have a few thousand dollars to spare and the inclination to do it. Hundreds of thousands of people have had their genome sequenced already.

A *few thousand* dollars? It cost about three *billion* US dollars for the Human Genome Project to produce the first-draft human genome sequence. It has been estimated that, in 2001, it would have been possible to sequence an individual human genome for about US$100,000,000. The cost has come down at a remarkable rate as the technology has advanced. Now, it's possible to sequence a human genome for (notionally) less than $1,000 ... and the cost is still falling. Analysing the data is already more of a challenge than generating the sequence. To put the fall in costs into context, imagine that sequencing a genome was a brand-new Lamborghini, retailing at $428,000. If Lamborghinis were to come down in price to the same degree that sequencing a genome has done, you'd now be able to pick up your shiny new car at the bargain basement price of $4.30.

Got a few bucks? Let's take this baby for a spin!

3

The boy who wasn't short

You will never make a crab walk straight.

ARISTOPHANES

Different people are prone to different types of mistake. I'm particularly vulnerable to the mistake that underlies much of magic. Magicians rely, in part, on misdirection — guiding your gaze over *there* so you don't notice something important that's happening right *here*. In medicine, misdirection can come from other doctors, from the patient, or just from unlucky happenstance, and it tends to lead to what look like simple mistakes — errors you would never make if you were paying attention to the right thing.

If the magician's art is to misdirect us, the art of medicine often lies in finding ways *not* to be misdirected. We talk about 'traps for young players' — but the truth is that old players can also be ensnared.

A few years ago, a general practitioner, not a young player, referred a small boy to me for investigation of short stature. This was a bit unusual, because most of the time such referrals go to a paediatrician first. Then they might go to an endocrinologist, a specialist in hormones, including those that direct growth. But there are many genetic conditions that can make a child short, so, although unusual, it wasn't an unreasonable thought to ask a geneticist's opinion. In this case, the story was quite worrying, because there had been a rapid crossing of centile lines.

Paediatricians track children's growth using centile charts. These are graphs that show normal growth patterns, with lines representing different percentiles. Three per cent of children are taller than the 97th percentile for height. A quarter of children are shorter than the 25th centile for height. Half of all children are lighter than the 50th centile for weight. And so on. Most children, most of the time, grow along a particular line. Start small, relative to other babies — you will probably continue to be small.

The neat thing about centile charts for growth is that you can very easily use them to track whether things are progressing normally or not. Does that baby have a big head purely because he's from a family who take large hat sizes, and is destined to do the same? Then he should track along the same centile line over time. Is his head crossing lines upwards? That might be a problem. Cross enough lines and he will most likely score a brain scan. Similarly, when someone has been on a particular track for height and then drops — like the boy in this story — that's a worry, and it makes us sit up and pay attention. Growing is one of the most important things that children do, and, when they stop doing it, it's important to find out why. It's not that this child had shrunk, of course — more that he had grown one centimetre over a period when we would have expected him to grow seven.

I went through the usual process we follow with any new patient. I asked about the boy's family and their heights. I found out about the pregnancy, about his birth and early development. I examined him, looking particularly for abnormal limbs, for disproportion between limbs and body. I looked at the creases on the palms of his hands, because, if the bones in your hands are short, the creases can form differently.

I found nothing. Not a thing. He was a completely normal-looking child, who to all intents and purposes had been doing fine — until his growth went off a cliff.

So I went back to the growth chart. Fortunately, his mother had brought in his 'Blue Book', a book new parents are given for recording health information about their child. Even more fortunately, there were several previous height measurements in the book. I plotted them on the growth chart — and the answer jumped out at me.

Every measurement throughout his life had placed this child a little below the 25th centile for height — except for the one that someone had done nine months earlier, which had him above the 90th centile. In retrospect, it was obvious that that measurement had been a mismeasurement: the boy had not plummeted from the 90th to below the 25th centile, because he had never been on the 90th in the first place.

The boy wasn't short, and certainly didn't need to see a geneticist. But I didn't count this appointment as a waste of time. And in the long run, neither did the boy's mother, because that referral may have saved her life.

When I was a medical student, cancer was a mysterious thing. Not that we were completely ignorant — far from it. We knew that there were plenty of things that could give you cancer. Smoking, of course. Asbestos. Certain viruses, such as HIV. Exposure to mustard gas, fortunately not a common problem nowadays. In Australia, melanoma heartland of the world (Come visit! You'll love it here!) ... the sun.

We even knew that there were some inherited types of cancer, and there was evidence as far back as the late 1950s that there were genetic changes in cancer cells. In particular, in 1959, two researchers in Philadelphia (Peter Nowell and David Hungerford) noticed that, in some leukaemia cells, chromosome 22 was abnormally short; it was named the Philadelphia chromosome. In 1973, Janet Rowley[1] discovered that the reason for the short

---

1   Just a few weeks after Rowley's discovery, Margaret Garson (a cytogeneticist in Melbourne) independently found the same thing. The two were friends, and the story told in Australia is that, when Rowley found out about Garson's findings, she graciously offered to publish their findings together; but Garson declined, saying that Rowley had beaten her to it and deserved the credit.

chromosome 22 was that part of the chromosome had broken off, and was stuck onto chromosome 9. This turned out to be enormously important, because it was the first of a whole class of chromosome abnormalities uncovered in cancer.

Many years later, the reason why this rearrangement of two chromosomes was a part of causing cancer was identified. The places where the two chromosomes break are in the middle of two different genes. Their fusion makes a new, hyperactive gene that drives abnormal cell growth. This discovery led in turn to the creation of a group of new treatments for some types of cancer (called tyrosine kinase inhibitors).

Over the past few decades, the biology of cancer has been worked out, in ever greater detail. It turns out that cancer is almost entirely a disease of the genome. The sickness that afflicts the genome of a cancer cell boils down to one thing: a mismatch between the accelerator and the brakes of the cell.

Growth is a fundamental part of life. At conception, you were a single cell. It was a huge cell, as cells go, about as wide as a strand of hair — but still a tiny, tiny thing. One of the most important tasks that cell had was to grow. As it divided and divided, signals were sent to the machinery of the new daughter cells, urging them to multiply and expand. These signals were obeyed, and, thanks to the rich bath of nutrients your mother provided, they were obeyed with gusto.

Which was fine — until it wasn't. At some point, it's not enough to be an ever-expanding ball of cells. You needed a shape. You needed some parts to grow, while others stopped. You even needed some cells to die. Six weeks after conception, you weighed about 500 times as much as when you were just a fertilised egg. If you'd continued growing at that rate, you'd have weighed more than the Earth before your first birthday.

This means that, to balance that first rush of acceleration, your cells also needed brakes. They needed a lot more than that, because all sorts of decisions had to be made. Which end is the top, and which is the bottom? Which side is the left, and which is the right? There are a pair of genes, named *LEFTY1* and

LEFTY2,[2] that hold part of the answer to that one. Once you have a top and a bottom and a left and a right, you also have a front and a back. You're still little more than a blob at this point, but you're on your way.

There's a beautiful complexity to what happens next. Proteins signal to each other in a kind of dance; instructions are sent to cells telling them their fate. *You*, and all your daughters, will be skin cells. *You* will be a nerve cell. *You* will be a liver cell. Grow along *this* line. Stop growing when you reach *this* point. Start doing your job, whatever it is: contracting, to make the heart beat; firing electrical signals, to make the brain work; filtering blood, to clean it and make urine.

But for some cells, the message reads: *you will die*. This is important in all sorts of ways — there is a process called programmed cell death, a pruning that takes out cells that aren't needed. One of the places this is easiest to imagine is in the limbs. Your arm started out as a little nub, then grew to be a flipper. Your hand started as a lump on the end of the flipper. To make fingers, the cells in the gaps between the fingers had to go away — and so they did.

The message telling a cell that it needs to die is important later in life, too. Cells that are sick or damaged can trigger their own destruction. If this didn't happen, a sick cell could cause problems by using up resources, by poisoning its neighbours, or by just getting in the way. Or by transforming into a cancer cell.

So there are three types of signal. Grow — the accelerators. Stop growing — the brakes. And the signal to die. All of these need to be in balance, and the balance is different for each cell type. In fact, for many individual cells, the brakes are locked fully on. Once a cell gets to be a mature white blood cell, for instance, by and large it is never supposed to divide again. Once it wears out, it has to be replaced by stem cells in your bone marrow. Stem cells hover around, not fully formed into

---

2   Yes, there should be a gene called *RIGHTY*. But there isn't. The closest I can offer is a gene that's important for the midline of your body, called *MID1*.

a specialised cell but waiting until they are needed — at which point they divide, and one of the daughter cells matures into the needed blood cell (or liver cell or muscle cell, as the case may be) and the other steps back into the wings, waiting until it is needed again. Some other kinds of mature cell (such as skin cells) keep the ability to divide and replace themselves.

In some parts of the body (cartilage, brain), cells last a long time and seldom need replacement. Elsewhere, there is a great deal of wear and tear. Skin cells and the lining of your gut are constantly shearing off and being replaced. Rapid growth and replacement of cells creates opportunities for things to go wrong, which is one of the reasons skin cancer and bowel cancer are so common.

Every time a cell divides, its DNA has to be copied. Two sets of three billion pieces of information, copied out in the space of a few hours, in a tiny, tiny space. Mistakes get made. They get made *all the time*. Mostly, they are caught in time by the genome's fact checkers, and are fixed. But there are plenty of chances for the fact checkers to get it wrong. A typical human body is made up of perhaps 30 trillion cells (30,000,000,000,000).[3] In the course of an average lifetime, that might mean around ten *quadrillion* cell divisions (10,000,000,000,000,000). That's a lot more than the number of cells, because of the many divisions needed to go from fertilised egg to adult, and because of all of the replacing of dead cells that has to happen.

---

3   That's just the *human* cells, of course — you are host to so many bacteria, protozoa, and fungi that there are about as many non-human as human cells in your body. Most of them are tiny, so at least the human part outweighs everything else. You probably woke up this morning thinking you were a human being, and you were only about 3 per cent wrong. Ninety-seven per cent human: not so bad.

It's not just single-celled organisms, by the way. When I was at university, we had a lecturer who claimed that we have so many worms in our guts that, instead of saying 'good morning' to people, we should say, 'How are your nematodes today?'

If that thought bothers you, I suggest you don't look up face mites. No, really — forget I ever mentioned them.

Sorry.

How many mistakes get missed? Well, if you're typical, you have between 40 and 80 changes in your DNA, in every cell of your body, that you didn't get from the genome of either parent, but will pass on to your children. Most of those had their origin in your father's sperm. Making sperm is a high-speed, high-volume, and comparatively low-quality exercise compared with the bespoke tailoring that goes into making an egg.

Do these new changes in the DNA matter? It depends on exactly where the mistake happens. It *can* matter a great deal, as we shall discover. But mostly the changes happen in places where it doesn't matter much: in between the genes, or in other places where it doesn't make a difference if you change a C to a T or an A to a G.

What about your body's allotted ten quadrillion cell divisions? How accurately is the DNA copied? Well, a recent study suggests that, typically, there is one new mistake per cell division. Yes, that's right — *almost every time a cell divides*, something goes wrong, and the genome is imperfectly copied. Your genome is decaying, every minute of every day. In a very real sense, you don't have 'a' genome — you have trillions of slightly different versions of the genome you started with. Almost no two cells share the exact same genetic make-up as each other.

If you have three billion bases of DNA, and there are ten quadrillion opportunities for it to be incorrectly copied, it follows that practically every single simple genetic change that is possible must happen many, many times over in a person's lifetime. Plenty of chances for the brake genes to be damaged and to stop working, or for accelerator genes to be jammed in the 'on' position. Or for completely new accelerator genes to be created, as when the Philadelphia chromosome is formed.

You might ask how any of us are alive; how is it possible to survive from conception to birth without having cancer, let alone through childhood and into adult life?

It turns out that there are a couple of different answers to that. Firstly, and most importantly, the transformation from healthy cell to cancer cell is not something that happens overnight, or (usually) as the result of a single genetic mistake. There are backup systems in place, so that, for the most part, a single error isn't enough to cause cancer — there need to be more piled on top of the first one, and of course they all have to happen in the very same cell. It's a bit tricky working out exactly how many mistakes it takes, because one of the features of cancer cells is that their DNA copying often gets sloppy, leading to mistakes being made. Some of these aren't part of why a cell is cancerous, but are a consequence of the chaos that goes along with being a cancer cell. This means that as well as 'driver' mutations, there are also 'passengers'[4] — not the cause of the problem, but along for the ride — and, when we study the DNA of cancer cells, it isn't always obvious who's doing what.

Thanks to our ability to sequence the whole genome of a tumour and compare it with healthy tissue, we are building up a picture of what it takes to go from being a well-behaved, functional cell to being a menace to society. It turns out that it's probably not a very large number of mistakes — often as few as six or seven, and for some types of cancer it can be even fewer, maybe just one or two. But remember that they have to be *exactly* the right (or wrong) combination of mistakes. Our bodies are full of almost-cancers that will sit there and never bother us.

How frightening is the problem? Well, if you are of a nervous disposition, you may want to skip this next bit.

They probably didn't think of it in quite this way, but, in 2015, a group of scientists from the United Kingdom set about working out exactly how scary the problem is. Four people who were having operations to tighten up sagging upper eyelids donated the excess skin for research. The skin samples looked completely normal under the microscope. The researchers in question took the leftover bits of skin, punched out 234 tiny biopsies, and read the genetic sequence in each of them. They

---

4   In this case, a 'driver' can be either a faulty brake or an overactive accelerator. It's anything that makes the cells grow faster.

looked only at a set of 74 genes that were known to be involved in the development of cancer, and within that they looked for changes in the genes that they could reasonably identify as potential drivers for cancer.

I don't know what they expected to find, but what they actually found was ... horrifying. Here's an image from the paper:[5]

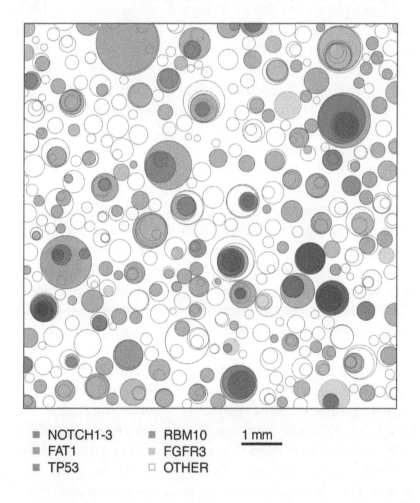

■ NOTCH1-3	▨ RBM10	1 mm
■ FAT1	▨ FGFR3	
■ TP53	▢ OTHER	

This picture represents one square centimetre of skin. Every circle is a region filled by cells that have a potential cancer-driver

5   From DOI: 10.1126/science.aaa6806. Reprinted with permission from AAAS.

mutation. Each circle would have started as one cell that multiplied and expanded until its daughter cells filled the space. Where there are overlaps, there are cells that have more than one driver mutation. Five key genes or sets of genes get their own colours, and the other 67 genes are represented by the unfilled circles.

Here's some consolation if that's freaking you out a bit (it certainly gave me the willies when I first read the paper). This was sun exposed skin, and the changes were mainly the sort of changes caused by ultraviolet radiation.[6] Also, the volunteers were aged 55–73. You wouldn't find anything like this picture in, say, a piece of muscle sampled from a 20-year-old. On the other hand, these were all Brits. You call that sun exposure?

So why *aren't* we all dead already? Firstly, you do need that series of unlucky breaks to happen to the *same* cell. Secondly, your immune system is pretty good at mopping up cancer cells — before any cancer can establish itself and get growing, it has to evade the cellular police. For this reason, people with immune deficiencies are at increased risk of getting cancer.

There's another group of people who are at increased risk of getting cancer. Some people are unlucky enough to inherit a faulty brake or accelerator gene from one of their parents. If the change is there in every single cell from the word go, potential cancers have a head start. And so we see bowel cancer families, breast and ovarian cancer families, and so on. In families like this, cancers tend to be diagnosed at much younger ages than usual — because the cancer cells have been brewing since much earlier. So when we see a family in which there are multiple people affected by cancers that we know can be associated with a particular gene (like breast and ovarian cancer are), and many of them are being diagnosed at a young age — we suspect there may be one of these conditions in the family.

There are a surprising number of different conditions like

---

6   Nowadays, I don't even turn on the lights without putting on sunscreen first. You can't be too careful.

this, but most of them are very rare. By far the most common involve a high risk of (mainly) bowel cancer and a high risk of (mainly) breast or ovarian cancer. It's not always as easy as you might think to spot families with such a condition. Cancer is common, and you can get what looks like a familial cancer syndrome just by bad luck. On the other hand, it's an increased risk, not a certainty, and there are many people who inherit one of these faulty genes but get away with it, living their whole lives unaffected by cancer. On top of that, men who inherit a mutation in one of the breast cancer genes, *BRCA1* or *BRCA2*, are much less likely than the women to get cancer. They *can* get breast cancer — far more frequently than the general population, but it's still uncommon. They have an increased risk of some other types of cancer, too, but the pattern is much less distinctive. All of this means that, sometimes, the family trees can be confusing and hard to sort out.

This was also one of the challenges faced by the people who were trying to find these genes in the first place. Nowadays, it's possible to make a link between a genetic condition and a specific gene by studying a remarkably small number of affected people. This is because we have access to the sequence of the human genome, and because of the power of the new genetic sequencing technologies. For example, my recent PhD student Emma Palmer is working on the epileptic encephalopathies, a group of severe neurological conditions that affect small babies. In the course of her PhD studies, she played a major role in discovering that faults in no fewer than *four* different genes can cause epileptic encephalopathy. Terrific researcher though Emma is, this would have been an inconceivable feat for one PhD student until the last few years.

In the late 1980s and through to the mid-1990s, when *BRCA1* and *BRCA2* were found, it was a long, hard slog making such a discovery. *BRCA1* and *BRCA2* are brake genes, and mutations in the genes lift the cells' feet (as it were) off the brake a little. It had been recognised for decades that some families have an inherited form of breast and ovarian cancer, with younger age at diagnosis on average than is usual for affected people in the general population. Over a number of years, researchers looked for genetic markers that could lead them to the genes. Think of the fly maps from the previous chapter, but, instead of finding that visible features like yellow body and white eyes were linked, the researchers were trying to find variable stretches of DNA that were connected to the occurrence of cancer. The closer they got to their targets, the tighter that linkage would be.

The process involved hundreds of researchers and many, many family members who volunteered as research subjects. In 1990, a group of researchers led by Mary-Claire King at the University of California, Berkeley announced that they had narrowed down the location of BRCA1 to chromosome 17. Gradually, her group and others whittled away at the chromosome. In May 1994, a group from the University of Utah, in collaboration with a group from Cambridge University in the UK, published a detailed map of the region, which by then contained only a little over 20 genes, one of which must be BRCA1. By October 1994, the race to find the gene was won ... but along with the announcement, and publication of the sequence of the gene, by Mark Skolnick and colleagues from the University of Utah, came a shock for those in the field. A company called Myriad Genetics, formed by Skolnick's group, had applied for a patent on the gene. This was one of a series of patents that essentially tied up commercial testing for BRCA1. And then they did it again. In a similar international effort, the location of BRCA2 had been narrowed down to chromosome 13. In December 1995, another British-led group, headed by Michael Stratton, published the sequence of BRCA2. But the day before the paper was published, Myriad announced that it, too, had found the gene, and had filed for a patent.

To most people in the field, the idea of patenting a gene seemed, and seems, absurd. Myriad didn't invent either gene, after all — they discovered it. Moreover, Myriad's discoveries would not have been possible without the work done to map the genes by many other, publicly funded groups, and all of that work rested on patients who volunteered to take part in research. It seemed deeply wrong to most of us that a company could profit from this. But profit they did, and for a long time.

This played out differently in different parts of the world. In the US, Myriad held a stranglehold on diagnostic testing for *BRCA1* and *BRCA2* for nearly 20 years, until finally the US Supreme Court struck down their key patents, which previously had been repeatedly upheld by lower courts. In Australia, a company called GTG had a patent on all of the genome that did not code for genes — i.e. most of it. Spoiler alert: GTG didn't invent that either — but they had still been granted a patent for it, and they had argued that Myriad had infringed this patent. The end result was a deal that left GTG holding the BRCA patents in Australia. This was highly controversial and never really enforced, but it was a big relief to the field when the High Court of Australia also struck down the patents, in 2015.

It has been said that women have double the number of relatives that men do. When we take a family history from a woman, we generally learn twice as much about the family as we would if we had spoken to her brother instead. Sometimes, this really matters. When I saw the boy who wasn't short, I took a family history from his mother, because this is something we always do. She knew that her first cousin, her uncle's daughter, had recently been diagnosed with breast cancer at the age of 50. She also knew about two others in the family who had died of breast cancer at a young age, and another who had died of ovarian cancer. She didn't know that breast and ovarian cancer could be inherited, so it didn't occur to her to seek advice, and, in any case, we usually start by testing someone who is affect-ed by cancer. Her cousin's doctors could have put the pieces together — if her cousin's father had known and passed on the information.

When I learned of the family history, I arranged for the cousin to be seen by a cancer genetics service. They tested her, and found that she carried a mutation in *BRCA1*. Later, the woman I had seen also decided to be tested, and found that she, too, had inherited the faulty copy of the gene. In one sense, of course, this was bad news for her — far better not to learn that you are at high risk of cancer. But this gave her options, which included having more extensive screening than she would otherwise have had, with a chance of detecting cancers early, meaning a better chance of cure. It also gave her the option of having risk-reducing surgery — removal of her ovaries and breasts before cancer had a chance to develop. A tough choice, but one that we know can reduce the chance that *BRCA1* mutation carriers will die from cancer. Several other women in the family had the opportunity to be tested as well. It's likely that sooner or later, someone in the family will have their life saved by this information.

Although I'm a doctor, Angelina Jolie has saved far more lives than I ever have or could hope to. When she told the story of learning that she, too, carried a *BRCA1* mutation, and of her choice to have surgery, our cancer genetics services were swamped with people who had considered their family history in a new light and sought testing. We called this the Jolie effect. As a result, our labs were flooded with samples, many of which tested positive. The rate of people having risk-reducing surgery, as Jolie had done, doubled in Australia and in the US, and probably in many other countries. There are people who are alive today only because of Jolie's decision to speak out.

The risks for people with *BRCA1* and *BRCA2* mutations are greatly increased compared with the general population, but there are no certainties. Some carriers live their lives unaffected by cancer, while others with the same genetic make-up die young. This means that making a choice to have surgery represents a decision in the face of uncertainty.

Offering people choices with uncertain outcomes is standard operating practice in genetics, as we shall see.

4

Uncertainty

Fate shuffles the cards and we play.

ARTHUR SCHOPENHAUER

Fate is not an eagle, it creeps like a rat.

ELIZABETH BOWEN

Jason couldn't remember a time when things had been okay at home. When he was a small boy, perhaps they had been. But then his father started to have violent mood swings; he would fly into a rage over nothing at all. There was never any violence, but Jason's parents fought constantly, for years. Eventually, when Jason was 12, his mother packed up their things and left, taking him and his sister to live in a different state. Jason never saw his father again.

Years later, thinking back on that time, Jason would wonder if perhaps his father had already been showing the first signs of the sickness that would, eventually, claim his life.

When I met Jason, he was in his early 30s. He and his partner, Lauren, had been living together for two years. They were thinking of marriage, of having children, of a future spent together. But first, Jason wanted to find out if he had a future at all.

Jason showed me a letter he had received eight years earlier.

Addressed 'to whom it may concern', the letter said that Jason's father had been diagnosed with Huntington disease, that this meant that his relatives were at risk of also developing the condition, and that the recipient might want to consider speaking to a clinical genetics unit about this information.

Huntington disease[1] is not rare by the standards of genetic conditions — it affects about 1 in 10,000 people. It's a cruel condition, with a characteristic trio of effects: a movement disorder, psychiatric problems, and a decline of mental function. There is a great deal of variation in the way that HD symptoms start and progress, but a typical story is of someone who had previously been well reaching his or her early 40s and starting to show signs of being affected. At first, these are subtle, easy to dismiss as by-products of normal ageing and entry to the middle years of life, rather than something more sinister: clumsiness, apathy, anxiety. Other people may start to notice that the affected person is having involuntary movements — twitching of the limbs and face. Depression may set in; balance and the ability to do complex physical tasks are lost. Over time, the brain's decline becomes worse and worse; eventually affected people lose the ability to care for themselves, to walk, to speak, and to swallow. All of this takes *years*; most survive a decade or even two after the first symptoms begin. It is a process of slow neurological demolition.

From the earliest descriptions in the first half of the 19th century, the familial nature of the condition was recognised. HD is inherited in a dominant fashion: if you're affected then each of your children have a 1 in 2 chance of inheriting the faulty gene and being at risk. There's something special about the fault

---

1   As we shall see in chapter 7, it's not at all unusual that George
     Huntington's 1872 description of the condition that has borne his name
     ever since was not the first such description ... or the second, or the third.
     Others had beaten him to it with publications in 1832, 1841 (and 1842),
     1846, 1860, and 1863. Five others could have had their names attached to
     the condition, but didn't. Three of them published their descriptions before
     Huntington was even born!

in the gene: HD is one of a class of conditions known as triplet repeat disorders. You'll remember the genetic code, made up of groups of three base pairs of DNA. In Huntington disease, the problem lies with a repetitive stretch of DNA: CAG CAG CAG CAG CAG ... This codes for the amino acid glutamine, so that in one stretch near the beginning of the protein you have glutamine-glutamine-glutamine-glutamine-glutamine ... and so on. Most of us have 35 or fewer repeats — typically 15 to 20 — and no chance of developing Huntington disease. People with 36 to 39 repeats *might* develop HD, sometimes starting later than usual, and progressing more slowly. With luck, they might never be affected at all. People with 40 or more repeats *will* develop HD, if they don't die of something else first.

There's a twist to this tale. If the repeated section is 27 repeats or longer, it's too much to cope with for the machinery in the cell that's responsible for copying DNA while making an egg or sperm. Mistakes can be made. In this case, it's a very specific type of mistake: the DNA copying mechanism has a risk of slipping — like a stuck record. CAG CAG CAG *skip* CAG CAG *where was I?* CAG CAG CAG ... This means there's a chance that the child will inherit a larger (or, sometimes, smaller) repeat than their parent.

For reasons we don't understand, it makes a difference whether a man or a woman passes the expansion on. On the whole, it's worse to inherit the HD repeat expansion from your father[2] than from your mother. The chance of a reduction in size is much less, and the chance of an expansion much more. Since larger repeat sizes lead to earlier development of symptoms, one of the consequences of all of this is that HD can actually get *worse* through the generations in a family — with earlier onset and quicker progression. There are a number of conditions that

---

2   You might think this is an example of the process for making an egg being superior to the process for making a sperm, but it seems like it's not that simple. For some other triplet repeat disorders, such as fragile X syndrome and myotonic dystrophy, it's the other way round — expansions *mainly* happen when the repeat is passed on by a woman.

do this; the phenomenon is called 'anticipation', despite the fact that early onset Huntington disease is hardly something to look forward to. It took a while to figure out this generational worsening, because the phenomenon is kind of weird and unexpected, and also because it's quite easy for family trees to give the *appearance* of anticipation just by chance. This is in part because the first person in a family who gets the diagnosis of a rare inherited condition is often a child with severe problems; that child's affected parent is probably more mildly affected. That looks like anticipation, and, although it isn't, you may need to see a fair few families before you can rule the possibility out.

You can probably see where things are headed here. If anticipation in Huntington disease means earlier and earlier onset through the generations, eventually it might get to a point when it starts *very* early — and it can. 'Juvenile' HD starts before the age of 20,[3] and children sometimes show symptoms when they are as young as five. The diagnosis of HD in such a young child is a double blow: almost certainly, his or her father has a smaller expansion, but one that means he, too, will develop the disease in time.

One more thing before you put yourself in Jason's shoes. For now, at least, Huntington disease has no cure. There are treatments that can lessen the symptoms, to be sure, but once the symptoms start, the condition is relentless, a slow-moving juggernaut that has only one destination.

Now ... let's say there is a 1 in 2 chance that this is *your* fate. We can do a test and give you an answer. Think you'd like to take the test?

For most people, it turns out, the answer is a very definite NO. Nearly nine out of ten people who know they are at risk choose not to find out. Their reasons vary. Many figure that, since there's nothing they could do about the information that

---

3 People with juvenile HD have at least 50 repeats; the youngest affected children can have more than 60.

will change the outcome, there's no real benefit to having the information. Some prefer uncertainty, and the possibility of a long and healthy life, over certainty that could be of a bad outcome. It's not knowledge you can un-know once you have it.

This means that the people who come to see a geneticist about testing are something of a select group. Reading this, you may think that you know what you would do, and that you are part of that select group, but it's quite possible that you are wrong. When people are asked whether, in principle, they would choose to be tested, about 80 per cent say yes — almost the reverse of what actually happens. This gap between what people say they would do and what they actually do when given a real-world choice has a name. It's called the intention–behaviour gap, and it applies well beyond the deeply personal and life-changing decision to have testing for HD. If you've ever signed up for a gym membership and then found yourself not, in fact, going to the gym — there's the gap.

Having said that, the group of people who *do* choose to be tested are generally very sure about that decision. Sometimes, like Jason, people wait for years before taking positive steps to have the test. But once you make it into my clinic room, it's very likely you'll go ahead — I can count on the fingers of one hand those I've seen who have reversed the decision at or after that point. When that does happen, it can be at the very last minute. Our department's files hold a sealed envelope containing the HD result of a man who had the test done but, when he was called with the news that his result was available, *changed his mind* about receiving it. For more than 15 years, that envelope has sat waiting, unopened. He has probably lived out the answer — developing symptoms, or not — by now.

From time to time, I do see people who come for testing but don't really want to know the answer. Mostly, they are in a situation that is similar to Jason's: they are planning a family and don't want to pass the condition on to their children. This opens up an unusual option: an *exclusion* test.

The gene we are interested in here, *HTT*, sits on chromosome 4.

The idea of exclusion testing for HD is to identify if an embryo has inherited its copy of chromosome 4 from Jason's mother, not his father. Remember that Jason has two copies of chromosome 4, one inherited from his father and one from his mother, and Jason's father had HD, but his mother did not. Jason will pass on either his father's chromosome 4 or his mother's, to each of his children.[4] If the embryo has its grandmother's chromosome 4, it's in the clear. The laboratory would make no effort to work out which of Jason's *father's* two copies of chromosome 4 harbours the faulty gene, because the information isn't needed to achieve the desired result. That way, we could be certain the baby won't have HD without revealing Jason's status.

Jason and Lauren considered this option, but Jason had already decided that he needed to know what would happen to him before making plans for the future. When I met him, it was his second visit — he'd already met with Lisa Bristowe, the genetic counsellor with whom I work. They had talked through the issues, the reasons to be tested or not to be, the possible implications for insurance,[5] and how Jason might deal with either of the possible results. Lisa had been sensitive to red flags: Is this someone for whom a bad result might be a truly devastating blow? Would he need extra psychological support around the time of the result? We had arranged for Jason to meet with a neurologist, who had found no signs of HD; this meant that, if the result were positive, it would not mean that Jason *had* HD, but that he was destined to develop it later. We'd

---

4   It's actually a little more complex than that, because of *recombination* — a process of creating a new, mix-and-match chromosome containing parts of each of a person's two copies of every chromosome. So really we are looking to identify an embryo with the grandmaternal chunk of chromosome 4 that contains the *HTT* gene, not the whole thing.

5   As you can imagine, life-insurance companies are not big fans of Huntington disease. As I write, in Australia there is a moratorium on insurance companies discriminating against people on the basis of genetic test results, for life-insurance policies of up to $500,000 as well as some other types of insurance. This is voluntary, and we don't know if it will be permanent.

offered an appointment with a psychologist as well, but Jason had declined this.

At the second meeting, we ran over some of the same ground, discussed medical issues, and arranged the test. Two tests, in fact: we always do predictive tests in duplicate, because of the consequences of getting things wrong, and because the biggest single cause of laboratory error is testing the wrong sample. Rare though it is for two individuals' blood tubes to be mixed up, sending two separate samples, which are tested independently, makes it nearly impossible for this to happen.

Six weeks later, we met the couple again. Walking to the outpatients clinic that morning, I carried a sealed envelope containing Jason's result; just before the appointment, when I knew that he had arrived at the hospital and was waiting to see us, I opened it.

I don't believe in luck, not really. But Jason had come to see me part way through a freakish two year period during which *nobody* that I tested for HD received bad news — and he did not break the run.

For Lisa and me — although not, of course, for Jason and Lauren — this was a pretty routine, straightforward scenario. But the ability to do 'predictive' testing can throw up situations that are not straightforward at all. Consider identical twins who have a 1 in 2 risk of developing HD. One wishes to be tested, the other does not. By testing one, you have tested the other. *We* would not communicate the information to the other twin, of course ... but what are the chances that she will inadvertently find out, or, to put it another way, how likely is it that the secret will be kept from her?

If, somehow, she is not told, how will she live with the knowledge that her twin knows the answer for both of them? Imagine being the twin who does not know, having an ordinary conversation with your twin and knowing all the time that she knows your shared fate — and could tell you the answer in a second, with as little as a nod or a shake of the head?

I've never encountered this situation, but from time to time we see people like Kim, a young man whose maternal grandmother had recently died of HD. His mother did not wish to be tested, but he did. Good news for him would tell us nothing about his mother[6] — but if he had inherited the faulty gene, then so had she. In that case, she accepted his decision to be tested, and it turned out that his 1 in 4 chance (1 in 2 chance his mother had inherited the faulty gene, times 1 in 2 chance she had then passed it to him) did not come up, leaving his mother where she had started.

What about testing children? It's natural for parents to be deeply concerned about the possibility that their children may develop a condition such as HD in the future, and to have a desire to find out — driven mostly, perhaps, by the hope of receiving good news. Very soon after predictive testing became possible, the genetics community decided that we should say no to such requests. There are various reasons for this: concerns of genetic discrimination, and of stigmatisation; worries that children will be treated differently, in a way that will harm them. The most compelling reason, for me, is that by testing them we take away their option *not* to be tested. Knowing that most adults, given the choice, do not have testing, is it fair to a child to take that future possibility off the table?

All of this hinges on the fact that we have no treatments yet that change the outcome in HD, although there is a great deal of research directed at developing such a treatment. If we knew that there was a medication that could prevent the disease from taking hold, and that you needed to start taking it when you were a child in order for it to work, the rules would change immediately. There are other conditions in which the stakes, and, as a result, the rules, are different in this way. Take familial

---

6   Almost nothing. There's a branch of mathematics called Bayesian
     probability, which is something of a favourite of geneticists. It allows us
     to combine different types of information to modify our assessment of
     the likelihood that something will happen. Without digging too much
     into the details, in this case, the finding that the man did not inherit an
     HD expansion shifts the probability that his mother had inherited an
     expansion from 1 in 2 to 1 in 3.

adenomatous polyposis (FAP), for instance. This is a condition in which hundreds to thousands of growths, called polyps, form in the colon. Left alone, colon cancer is inevitable; surgery to remove the colon is needed once the polyps are present. Screening with colonoscopies is needed to see if that has happened yet, starting from around 10–12 years of age. Polyps tend to start showing up in the mid-teens, but they can appear even earlier.

Because of this, genetic testing in a child who is at risk of having FAP is completely uncontroversial, although not taken lightly. There are important differences between FAP and HD. The age at which problems develop is generally much earlier in FAP — genetic testing in a child would not usually be happening decades before the first symptoms might appear. And you can *do* something about it, using screening and surgery ... and the something that you can do is relatively burdensome,[7] adding an incentive to do testing, so that you can spare half of the at-risk kids from having to go through it.

Falling somewhere in between these extremes is a group of conditions that affect the heart. A cardiomyopathy is a disease of the heart muscle. Most commonly, the muscle becomes thickened,[8] which can block the flow of blood through the heart; or it becomes weak and floppy.[9] Either type can lead to problems with the flow of electricity through the heart, with potentially fatal results. Other conditions, such as long QT syndrome, affect that flow of electricity without changing the heart muscle.

These are mainly dominant conditions, like HD and FAP, and they vary enormously within and between families affected by them. I once saw a 14-year-old boy who popped into a fast-food restaurant on his way home from school. It happened that the woman standing in line behind him was a nurse. This was

---

7   I can attest from personal experience that having a colonoscopy is no big deal, thanks to the wonders of anaesthesia — but preparing for one (bowel washout) is not fun at all.

8   Hypertrophic cardiomyopathy.

9   Dilated cardiomyopathy. There are several other types of cardiomyopathy that are much less common than these two.

fortunate because, when the boy's heart suddenly stopped, the nurse performed CPR until the ambulance arrived. Remarkably, the boy suffered no ill effects, but it turned out that he had quite a severe cardiomyopathy that had caused his cardiac arrest. We were able to identify the genetic cause of his condition and track it through the family. There were several people in the family who turned out to have relatively mild heart problems — including the boy's mother. But *her* father, who was in his 70s, carried the same genetic change as his grandson, and had done so all his life, yet his heart remained healthy.

For adults, at least, deciding to be tested for conditions like this is generally much more straightforward than the decision to be tested for Huntington disease. The implications are worrying, to be sure, and it might be distressing to find out that you are at risk of developing a serious heart condition. But there are treatments that can change the likelihood that you'll die from the condition, and there's that chance that you might never develop symptoms. Still, though — should we test children in families like this, to see if they have inherited the at-risk version of the gene? What would the result mean? It's not 'predictive' if you can live your whole life without ever developing a problem. On the other hand, these are not always adult-onset conditions — quite young children can be affected sometimes. There are treatments that can reduce the risks ... but you could also just screen for problems using non-genetic tests such as echocardiography (an ultrasound of the heart), which are not invasive or burdensome, and then treat any problems as they arise.

It's not at all clear what we should do about this,[10] and different geneticists have come to different conclusions. My own position has shifted over the years; now, I explain all the issues to parents who want to have their children tested, I make sure they understand the implications, and then, if they still want to go ahead, I do the test. If the child is old enough, they get to

---

10   It's not clear to geneticists. Cardiologists are generally very clear-cut about the issue: they want us to just get on with it, so that they don't have to screen people who don't need to be screened.

have a voice in the decision-making process. Teenagers often say no.

The implications of the results are potentially a bit different for a heart condition than for a brain condition or cancer. Probably the risk of stigma isn't there, but there are some extra concerns. Want to be an airline pilot when you grow up? How sympathetic is the company doctor likely to be to the idea that you have a genetic test result that says your heart might suddenly stop one day? Never mind that you might also live a long and healthy life and never show effects — aviation is a risk-averse business, and it wouldn't be at all surprising if this type of information turned out to be career-limiting.

In HD, FAP, and inherited heart conditions, the uncertainty is often about *whether* to have a test. But sometimes, uncertainty follows from the result, rather than preceding the test.

Lee-Ann and Derek had been trying for a baby for what seemed like a long, long time. Test after test had not found a cause for their infertility, beyond one that had been obvious from the beginning: time was not on their side. When they first went to see a doctor about their difficulties conceiving, Lee-Ann was 37 and Derek was 40; by the time I met them, she was 41, and pregnant at last, a naturally conceived pregnancy after several years of unsuccessful IVF. They told me that they had been overjoyed by the pregnancy, but worried about the possibility of a chromosome problem in the baby. Then, an early ultrasound scan showed an excess of fluid at the back of the baby's neck. Chromosomal conditions are among the possible causes for this finding,[11] so they chose to have chorionic villus sampling (CVS) done. This test takes a tiny piece of the placenta for use in genetic testing, with the idea being that, since the placenta shares its

---

11   There's a long list of other possible causes, but if the chromosomes and 18-to-20-week ultrasound are normal, the outcome is usually a healthy baby.

genetic make-up with the baby, if the baby has a chromosomal problem, it will be present in the placenta as well.[12]

Lee-Ann's CVS showed that, instead of the usual 46 chromosomes, there were 45; instead of XX or XY, there was just a single X. This was the point at which I was asked to see them, to talk about what this result might mean for their baby.

In general — and there certainly are exceptions — if you have two X chromosomes, you'll be a girl.[13] In general — and there certainly are exceptions — if you have one X and one Y chromosome, you'll be a boy. This is why the X and Y are called the 'sex chromosomes'.

When its development begins, every embryo carries the potential to be either male or female. Early structures called the Wolffian duct, which has the ability to develop into male sex organs, and the Müllerian duct, which has the ability to develop into female sex organs, are present in every embryo. If, at around six weeks after conception, the SRY[14] gene is activated, the Müllerian duct will wither and the Wolffian duct will develop. It's a boy! As its name suggests, SRY is on the Y chromosome. If there is no SRY signal — for example, because there is no Y chromosome — a different set of signals kick in. The Wolffian duct withers, the Müllerian duct develops. It's a girl!

Except when it isn't.

This is a complex process, and, like all complex processes, it has vulnerabilities — ways that things can go wrong. The 'disorders of sexual differentiation' (DSD)[15] include a spectrum of variations on a theme, ranging all the way from a boy with two X chromosomes to a girl with an X and a Y. Infertility is

---

12 Sometimes, there can be changes that are present in the placenta but not in the baby. More on this in chapter 11.

13 Here — and elsewhere in the book — when I refer to girls and boys, and male and female, I am referring to physiological sex rather than gender identity.

14 SRY, for 'Sex-determining Region on the Y chromosome'.

15 Also known as 'disorders of sex development' and 'differences of sex development'.

often part of the picture, and sometimes there are other medical complications, because some of the genes involved are important not just for sex development but for other parts of the body as well. Sometimes in these conditions, the baby's genitalia can be 'ambiguous' — when the midwife checks the baby, the best answer to the question 'Is it a boy or a girl?' may be 'I don't know'.

Perhaps surprisingly, abnormalities involving the sex chromosomes are only rarely a cause of any doubt about the sex of the baby,[16] although they can and do cause problems, in ways that relate to the special status of the X and Y. As discussed in chapter 1, for all of your chromosomes but that one pair, you really, really need to have two copies, and only two.

But the X and Y are special. Or, arguably, the X is special and the Y just basks in its glory. The X chromosome is large, and packed with important genes, including many that are essential for the way the brain develops, and thus for intelligence. The Y chromosome is mostly junk. Its genes include *SRY*, a handful of genes needed to grow testicles and make sperm ... and not much else.

Which leads to something of a mystery. Why is it okay for most men to have only a single copy of such a large, important chromosome — if having one copy of chromosome 21, a much smaller and less important chromosome, is uniformly lethal? Alternately, why is it okay for women to have two copies of the X chromosome if the single copy that men have is the correct number? Why isn't the extra chromosome causing problems in women?

The answer to this question was proposed by Mary F. Lyon, a mouse geneticist. Lyon had completed her PhD in the 1940s

---

16   The main scenario in which this sometimes happens is when the baby is a mosaic, with some cells having a Y and others not. Most commonly, this is a mixture of cells with 46 chromosomes, including X and Y, and other cells that have lost the Y and have 45 chromosomes with just the one X. Even in this situation, the most common outcome is a boy, although anything is possible — including a girl with Turner syndrome or a baby with ambiguous genitalia.

in Cambridge, under the supervision of R.A. Fisher, working on genetic mapping in mice.[17] In the years following World War II, she went on to study the chromosomes of mice that had been exposed to radiation, in work funded because of concerns about the possible effects of nuclear weapons on chromosomes. In a letter to *Nature* published in 1961, Lyon put together several previous observations, including coat colour patterns in mice with mutations that had been linked to the X chromosome,[18] and the fact that mice with only a single X chromosome were apparently normal females. She deduced that, early in embryonic life, one of the two copies of the X chromosome in each of the cells of a female mammal is switched off — inactivated — and the other is active. If a female mouse has a variation in a coat-colour gene on one of its two X chromosomes, and the normal,[19] or 'wild type', version of the gene on the other copy of the X chromosome, you can expect exactly what had already been observed in at least six different types of mice: a patchy pattern, with a mix of normal and mutant coat colours. Hair roots with the normal gene active will produce fur of one colour, and those with the mutant gene active will produce fur of a different colour.

Lyon's proposal became known as the Lyon hypothesis, and the process was named Lyonisation[20] (although it is now more commonly called, prosaically, X-inactivation). Every single

---

17   We'll meet Fisher again in chapter 9. In an oral history interview recorded in 2004, Lyon made it clear that Fisher, a famous theoretician, was no great shakes as an experimental geneticist. All the same, supervising Mary Lyon's PhD was arguably a major contribution to the field by Fisher, so his efforts in the lab were not wasted.

18   Just as there are tortoiseshell cats, there are tortoiseshell mice — and the reason, Lyonisation, is exactly the same for both species. There don't seem to be any tortoiseshell humans, which is a pity.

19   Of course, you can't really have an 'abnormal' coat colour, but the principle applies to X-linked genetic conditions.

20   The word 'lionised', as in 'lionised by the press', isn't so commonly used any more. But thanks to Mary Lyon, whenever I do come across it, I experience a moment of confusion, sometimes accompanied by an odd mental image of a tortoiseshell politician.

prediction Lyon made about the underlying biology has since been proved correct. Lyon recognised that her hypothesis might have implications for human genetics, but, at the time, there wasn't much known about human X-linked conditions. Now we know about lots of them. While males are often severely affected by such conditions, sometimes there are no effects in females at all; or effects can be patchy, whether in skin or in another tissue. At one extreme are conditions that are so severe that very few or even no males are ever seen — you need to have one functioning copy of the X chromosome to survive.

There are two parts of the X chromosome that are not affected by X-inactivation — they are switched on in both copies of a woman's X chromosome, and they have exact copies in the Y chromosome. These pseudo-autosomal regions (so called because they behave as if they are on one of the non-sex chromosomes, the autosomes) are at the tips of the chromosomes and are necessary for the sex chromosomes to behave normally in cell division during sperm formation in males. They aren't terribly large — PAR1, on the tip of the short arm of the X and Y chromosomes, contains just 16 genes, and PAR2,[21] at the other end of the chromosomes, contains just three. But that doesn't mean they aren't important.

If the PARs were just padding at the tips of the chromosomes, it probably wouldn't much matter how many X chromosomes you had, because the extra ones would be switched off. It also wouldn't matter if you had just a single X chromosome. And it probably wouldn't matter a great deal if you had multiple copies of the Y chromosome.

As it turns out, though, having an extra sex chromosome is not altogether harmless. Women with an extra X, so that they have 47 chromosomes with three Xs, tend to be tall for their families but otherwise are *mostly* just healthy people without particular medical problems. It's entirely possible to live your

---

21   Some people reportedly have a PAR3 as well! But since most of us don't, it probably isn't very important.

whole life with XXX and never know about it, or need to know. On the other hand, it is pretty clear now that, compared with women with just two copies of the X, those with XXX have a tendency to learning difficulties, and may have behaviour problems in childhood, or even autism. Having an extra Y (XYY) seems to have quite similar effects: most such men lead normal lives and never find out about their extra chromosome. XXY is more likely to cause noticeable problems — boys and men with the associated condition, Klinefelter syndrome, don't produce as much testosterone as usual, are more likely to struggle at school, and are infertile. Some of these problems are treatable with testosterone. Adding more chromosomes into the mix — 48,XXXY or 49,XXXXY for instance — does make things worse, as you might expect.

Even though it's decades since these conditions were first described, we don't have rock-solid information about all this, because of the way that most people with these differences in their chromosomes are identified in the first place. If you are doing fine, nobody is going to bother counting your chromosomes. This means that the people who have a chromosome test are not representative of the group as a whole — they are skewed towards the more severe end of a spectrum. This is known as ascertainment bias and is the bugbear of researchers studying any condition that can be mild or severe.

During the 1960s and 1970s, there were several studies of prison inmates that seemed to show that men with XYY were more likely to be incarcerated, and the idea that the extra chromosome made you more aggressive and thus more likely to be locked up persisted for decades. As recently as 2006, the question was still open enough that a group of Danish researchers conducted a large study that concluded that criminal convictions were indeed more likely in men with either XYY or XXY ... but that this effect almost entirely went away when adjusted for poverty, which itself is linked to a higher likelihood of being convicted of a crime. Since they were studying only men who were known to have an extra sex chromosome, and since these people were tested for a reason, it seems very likely that ascertainment bias is enough to explain the increased chance of criminal conviction that still remained after adjusting for poverty.

There have been a couple of studies in which large numbers of babies were screened for extra chromosomes and those with sex chromosome differences were then followed over time — a heroic effort, considering how many years you need to wait before you really know how things have turned out. As you might expect, the problems seen in those groups are generally much milder than those seen in people who were diagnosed because a doctor thought they had a problem that might be caused by a chromosomal condition.

Lee-Ann and Derek didn't need to know about the effects of an *extra* chromosome, because their baby was one short. In general, missing chromosome material is more of a problem than extra, and this situation was no exception. It's thought that 99 per cent of all babies that are conceived with a single X chromosome are miscarried, often before a pregnancy is even recognised. A common problem in pregnancy is a build-up of fluid in the tissues of the developing fetus, and this can by itself be severe enough to cause a miscarriage. Girls with just one X chromosome who get to be born have a condition called Turner syndrome. The effects of this are very variable indeed. Girls with Turner are pretty much always short,[22] and pretty much always infertile, but everything else about the condition is unpredictable.

Just how variable are we talking about? Well, a girl with Turner syndrome might be born with congenital heart disease and kidney malformations. Her neck might be webbed, and she may have a distinctive facial appearance. She might have significant difficulties at school — her intelligence would usually be normal, but there are some particular areas that she might struggle with, to the point of needing extra help. She might be

---

22   Growth hormone can help with this.

shy, anxious, and reserved. Planning and decision-making might be particular weaknesses for her.

Or she might have only some of those problems; or she could be completely fine, apart from being short. She might grow up not suspecting that she had any kind of medical problem at all, and the diagnosis might be made only when she is having tests to find out why she is infertile.

When I met Lee-Ann and Derek, I explained all of this, including the statistics about the likelihood of different aspects of the condition affecting a baby diagnosed in this way, as well as the weaknesses in the way those statistics were collected. I also explained that the only parts of all this that we could give them extra confidence about were the heart and kidney abnormalities, which mostly should be detectable on ultrasound. The rest was unpredictable; they would be facing uncertainty, much of which would take years to resolve.

For Lee-Ann and Derek, the decision about what to do with this information was relatively simple. They told me that if the baby had had a lethal problem, they would have requested a termination of pregnancy, and that if the baby had had a condition like Down syndrome, it would have been a challenging decision for them, and they weren't sure what they would have done. For them, though, Turner syndrome — even at the end of the spectrum with the most problems — seemed like a milder issue. They were sad about the prospect that their daughter would face extra challenges in her life, and the infertility that is part of the condition was particularly a blow to Lee-Ann. But they definitely wanted to continue with the pregnancy.

This decision is not as straightforward for most of the people I meet in similar situations. Many really struggle with the information, and, in the end, a majority of couples request termination of pregnancy after learning their baby has Turner or Klinefelter syndrome. Even when the finding is XXX or XYY, it's not uncommon to choose termination.

In dealing with this kind of finding, the uncertainty about what will happen is often one of the hardest parts.

All of us have to make decisions in situations where the outcome is unknown — decisions about relationships, about education and careers. There's a whole genre of self-help books about dealing with uncertainty, which speaks to how hard it can be. It's true that every prospective parent faces uncertainties about what the child will be like, and how they will cope with parenthood. But few are presented with such a specific range of possibilities and asked to make a tough choice, knowing that either decision will have a huge impact on their lives. To make things harder, from the time of the result, there is pressure to decide *soon*, because, if you don't, the pregnancy will be too far along and the choice will be out of your hands.

Uncertainty is a constant in clinical genetics, and it comes in various forms. Our stock in trade is situations like Jason's, in which the possible outcomes are clear, and the uncertainty is about whether to have a test; or like Lee-Ann and Derek's, in which the test has already been done, and the diagnosis is clear, but exactly what it will mean remains uncertain.

Increasingly, we find ourselves having to cope with another type of uncertainty. All too often, we do a test and find ourselves uncertain about whether the result means anything at all.

5

Needles in stacks of needles

… as we know, there are known knowns; there are
things we know we know. We also know there are
known unknowns; that is to say we know there are some
things we do not know. But there are also unknown
unknowns — the ones we don't know we don't know.

DONALD RUMSFELD

I've always felt Donald Rumsfeld was unfairly pilloried for the
comment above, because it seems like a pretty fair summary of
the way the world works, particularly the world of medicine.
I might further divide Rumsfeld's last category, the 'unknown
unknowns' into two: things we do not know about at all, and
things we *think* we know, but are wrong about. Worrying about
that last category is one of the many things that keep me up at
night.

In 2011, when I had already been working for more than a
decade as a clinical geneticist, a casual conversation with my
friend and mentor Michael Buckley turned unexpectedly into

one of those forks in the road we sometimes encounter in life. Michael is one of Australia's leading genetic pathologists — a doctor whose specialty is the oversight of the laboratory aspects of genetic testing. His lab, in which I now work, is an important centre for the diagnosis of rare genetic conditions. I mentioned to Michael that I sometimes wished that I, too, was a genetic pathologist; and he said, not really expecting me to take it seriously, that there was no reason I should not become one. The idea caught hold, and I wound up spending several years training part-time in laboratory medicine and sitting a series of exams to qualify as a pathologist. Now, I divide my time between the clinic and the lab, both ordering genetic tests on patients I have seen and writing reports on tests ordered by others and performed in our lab.[1]

Chance event though this was, the timing could not possibly have been better. When I started training, it was already apparent that a new type of genetic testing was on its way. Over the course of the next few years, that promise became a reality, leading to a transformation in the specialty. By sheer good luck, I've found myself in the thick of it.

You'll remember the extraordinary fall in the costs of sequencing an entire human genome, from billions of dollars to a thousand or so. There were some noteworthy waypoints on that journey from moonshot to utility. Craig Venter and James Watson were the first two named human beings to have their whole genomes sequenced. The third was a businessman, Dan Stoicescu. Dr Stoicescu, who made his money in biotechnology, spent some of it to have his whole genome sequenced by a company called Knome. They charged him US$350,000 for the privilege, which must have seemed like a bargain, considering that, just the previous year, it had cost three times as much to sequence Watson's. By the following year, Knome were charging

---

1   It's considered unwise to report on complex tests that you've ordered
    yourself, on the grounds that you may bring biases to the task that could
    cause you to miss unexpected findings, or over-emphasize results that fit
    with your preconceptions about the case.

just US$100,000 for the service: at that point, you would have needed to be very brave or very unconcerned about your money to shell out so much, for a product that was getting cheaper so very quickly.

Less spectacular, but arguably more important, was Knome's exome service. The exome is the 1–2 per cent of the genome that codes for protein: all the exons of all the genes, plus a bit on either side of each of them. Exome sequencing has the advantage that it's much cheaper to read the sequence of 2 per cent of the genome than the whole thing, and, because there's not much we can interpret that isn't in the exome, you don't lose much diagnostic capacity by only sequencing the exome.

On 5 October 2009, a scientist called Daniel MacArthur wrote an article in *Wired* magazine about the launch of Knome's US$24,500 exome service. Just five years later, MacArthur would become famous throughout the world of genetics for his leadership of the ExAC project, a collection of exome data from more than 60,000 people. ExAC, later reincarnated as an even larger database called gnomAD, has become arguably the single most important tool for interpreting the results of genetic testing.

Back in 2009, however, exome sequencing still seemed a long way from reaching the clinic. At those prices, it remained the province of rich individuals and very well-resourced research laboratories. Even when it seemed obvious that this would become a diagnostic test at some point, it was hard to imagine how *soon* it would happen.

The driver behind all this was technology. The Human Genome Project had relied on Sanger sequencing, amplifying and then reading the sequence of short stretches of DNA — typically a few hundred bases at a time. This works very well if your target isn't too big. A typical gene might have 10–20 exons. Amplifying up 10–20 small stretches of DNA, then sequencing that amplified DNA, and comparing the sequence you find with the sequence you expect is a fair bit of work, but it's achievable. It's a bit like being presented with a large book but being asked to proofread only the chapter titles. You *can* even sequence the entire genome that way (proofreading the whole book). That's how it was done first time round, after all — but you're talking about billions of dollars and years of work. That's true even today, if someone were mad enough to try it. Even attempting exome sequencing on an individual human using the old technology would be an impossibly daunting task: 300,000 separate stretches of DNA would need to be amplified, sequenced, and read.

It's obvious, then, that if you want to turn large-scale sequencing into an accessible test, you have to approach the problem in a different way. Right now, there are at least half a dozen different technologies that can do this, using a variety of different chemical tricks to do the job. At their core is a single unifying concept: reading many, many individual strands of DNA all at the same time. It's known as *massively parallel* sequencing, and the first to come up with a way to do it was a now-defunct company called 454 Life Sciences. The name is a bit of a mystery: there's a rumour that the company's original street address was at number 454, but it has also been suggested that this is the temperature (in Fahrenheit) at which money burns.

The company was founded by Jonathan Rothberg, one of the great inventors and entrepreneurs of the modern age of genetics. While still a student, Rothberg founded one of the first genomics companies, CuraGen, which was the parent company of 454. He went on to found two other important genomics companies, (the much better named) RainDance and Ion Torrent, among many others.

Rothberg's original motivation to get genetic sequencing going faster was the experience of one of his children being seriously ill in the newborn period. Rothberg thought that his child's doctors ought to be able to do a rapid genetic test to make sure that babies like his didn't have a genetic condition. He can fairly be said to have achieved that goal — the lab I work in uses an Ion Torrent sequencer, called a Proton, to do rapid exome sequencing to diagnose genetic conditions in sick babies.

The 454 company had a number of notable successes in the few years before the development of newer, faster, cheaper sequencers rendered their machines obsolete. James Watson's genome was sequenced by 454 Life Sciences; and Svante Pääbo, the evolutionary geneticist, used 454 sequencing for his first draft of the Neanderthal genome. This was the work that revealed that, in a way, Neanderthals are not completely extinct — thanks to interbreeding, most human genomes are a few per cent Neanderthal, and about a fifth of the whole Neanderthal genome lives on in the genomes of modern humans.[2]

To sequence the genome of a Neanderthal, you need to be able to work with tiny amounts of DNA — the little that has been preserved over the millennia. That DNA has been fragmented and degraded and is at risk of contamination by even miniscule amounts of modern human DNA. Sequencing such a genome is very different from working with the genome of a living creature that can keep on making more DNA to harvest.

Which brings us to one of the great unsung heroes of modern genetics — I'm referring to none other than NA12878. That may not seem like much of a name, but it's one that's famous among laboratory geneticists, because it's the code for the Genome In A Bottle. In truth, there are numerous Genomes in Bottles, not just one, but NA12878 is far and away the most famous and widely used. This code refers to DNA from a woman who lived in Utah in 1980. We know that, back then, her parents were still alive and she had 11 children (six sons and five daughters). She and her parents consented to extensive use of their DNA; consent was also given for her children (we don't know if they were old enough to consent for themselves at the time). Some of her cells were grown in the lab in a way that generated an essentially immortal line of cells, and huge quantities of DNA have been

---

2   We've since learned that our genomes hold traces of other ancient breeds of humans, including the Denisovans, named for the Russian cave in which a finger bone and a tooth were discovered in 2008.

harvested from those cells.

That DNA has had its genome sequenced over … and over … and over again. Practically everything that we can know about a person's genome is known about NA12878. As a result, she has become the gold standard. Just as all metric measurements used to trace back to a standard kilogram and a standard metre kept in sealed containers in Paris, so virtually all of the world's genomic laboratories refer back to this one woman's genome as their benchmark. You can buy tubes of NA12878 DNA — hence 'Genome In A Bottle' — to use as standard material. The lab I work in sequences her exome twice each month as a quality measure, to make sure the accuracy of our sequencing remains high. If we sequence NA12878, we know exactly what we *should* find, at every position in the genome; any differences from the known sequence must be errors. If Watson, Venter, and Stoicescu were the first, second, and third humans ever to have their genomes sequenced, NA12878 is undoubtedly the *most* sequenced human in the world, by a huge margin. When her 'name' comes up, I often wonder if she is still alive. It's 40 years since that simple act of altruism, when she gave a sample of blood for a research project. If she's still around — does she know how important she is?

Over the past decade, the applications of the new sequencing technology have moved from the realm of science fiction through to established reality and now to routine clinical testing. The impact this has had on genetics has been transformative — it has been truly wonderful to be around to see it. As a clinician, I spent many years seeing children with intellectual disability or other complex medical problems that might possibly have a genetic cause. Occasionally, we would make a diagnosis based on a pattern of features. Most often, though, we would do the tests available to us, think hard, consult databases or maybe even the Dysmorphology Club (see chapter 7) … and still come up empty-handed.

As a result, there was a whole branch of the genetics medical literature devoted to empiric recurrence risks. The idea of this was to look at families in which a child was affected by a condition, then see what happened to other children born in the family — simply counting the number of affected and unaffected children to get a percentage. Then, if we saw a child with that condition, we could use those figures to estimate the likelihood that a future brother or sister would also be affected. For intellectual disability, the figures varied from study to study but clustered around the 5–10 per cent mark. A 10 per cent chance that another child will have a severe problem is about as difficult as it gets for most couples who are trying to decide whether to have another child: it's a figure that's not very high, but also not really low. Do you take your chances? You might be waiting a long time before you know whether your next baby also has intellectual disability.

Now, our ability to make a diagnosis has dramatically improved — not because we are any better at our jobs, but because we have new and vastly improved tools to work with. If you consider children who had already had a chromosome test, we used to diagnose maybe one child in 20 with intellectual disability using the old approach. Now, for the more severely affected children at least, it's more like 50 per cent, and there are some groups for whom we can do even better than that. Better still, it's really common that the child's condition is due to a 'de novo' change in a gene — one that wasn't present in either parent. This is good news because it means a low chance that other children will have the same problem.

That chance isn't zero, because of a phenomenon called mosaicism. A mosaic tiled floor has a mixture of different-coloured tiles; similarly, someone who is a mosaic for a change in a gene has a mixture of cells, some with the change and some without. As we saw in chapter 3, in one sense all of us are mosaics, because of the mistakes that happen during cell division. Usually, this has no obvious effect, except for the tiny fraction of such mistakes that lead to the development of cancer. However, a change that happens in the first few cell divisions after conception might wind up being present in a substantial fraction of a person's cells, and can sometimes cause features of a genetic condition. These are often less severe than if the change is present in every cell, and can be confined to just one part of the body. If the condition affects the skin, it may be possible to tell that a person is a mosaic just by looking at them. In that case, we often see streaks of skin that look different, in a distinctive swirling pattern that follows the lines of Blaschko — the lines the dividing cells migrate along during the development of the embryo.[3]

A change that happens a little later after conception might wind up being confined to just a patch of cells, somewhere in the body. If a parent has a patch of cells like this in their testicles or ovaries, they can make more than one sperm or egg that has the change, and thus have more than one child affected by the same condition, despite having no sign of the change when we test their blood. This situation is called gonadal mosaicism: a mosaic state present in the gonads (the testicles or ovaries). The chance of a second affected child being born might be quite low — if most cells in the gonad have two normal copies of the gene — or might be as high as if the change were in every cell of the parent's body.

In practice, it is rare for more than one child in a family to be born with a genetic condition due to gonadal mosaicism — I've only seen it happen a handful of times — but it means we can't ever be completely reassuring that there couldn't be another affected child, even when the first child's problem is due to a genetic change that we can't detect in the parents.

Nevertheless, you might wonder why, if there were genetic diagnoses waiting to be made in so many of our patients, we

---

3   You can also think of everyone who has two X chromosomes (including most women) as being mosaic, because any variation in a gene on one of the two X chromosomes will be expressed only in the cells that have that copy of the X switched on. As a result, there are some X-linked conditions in which affected women have skin changes in a pattern that follows the lines of Blaschko. One such condition is called Goltz syndrome; another is the evocatively named incontinentia pigmenti.

were so lousy at making them in the past. There are a few different reasons for this. Some of the conditions we are seeing today were only identified recently, because of the availability of exome sequencing. The pace of discovery is so fast that, if we do exome sequencing and don't get an answer, one of the best follow-up tests we can do is just to come back and reanalyse the original data after a year or so. Surprisingly often, new discoveries made in that time mean that we can interpret data that we couldn't understand first time round, and we can make a diagnosis.

Some conditions were thought to be very rare indeed, but have turned out to be much more common than we thought — but also more variable, so that most cases are hard to recognise. And some conditions are just extremely rare; it's impossible for any doctor to know about all of the thousands of rare conditions out there, and the diagnostic databases we consult are incomplete and imperfect.

At the moment, most of the time, we use the new technology either for exome sequencing or to look at a more limited, targeted list of genes — the latter is called a gene panel. If you know there are only ten genes linked to a particular condition, it might not be worth your while sequencing more than 20,000. Sometimes, that's exactly what we do: we sequence every gene and then ignore almost all of the data, analysing only the parts we are interested in (I tell the story of a time we did this in chapter 10). It's already clear that, within a few years, when the costs come down a bit more, we will stop bothering with exome sequencing, and perhaps we will abandon panels, too. Instead, we'll just do whole genome sequencing, which is already a bit better than exome sequencing at giving us answers, and is likely to improve further. At that point, we'll probably also stop doing many of the chromosome tests we currently do, since the same information — in greater detail — will be available from the genome.

So that's all great: everything in the garden is roses. But — you knew there was a 'but' coming, right? — there are some problems we have to contend with. The biggest of these is dealing with the unknown.

The paper describing James Watson's genome, published in *Nature* in 2008, contained a clear signal that things were not always going to be straightforward. The researchers who studied Watson's genome saw what seemed like an anomaly, and had a stab at explaining it. With the benefit of that powerful tool the retrospectoscope, I can tell you that their explanation was completely wrong.

The clue was this: Watson was found to be a carrier of *ten* different genetic changes that had been reported as disease-causing in autosomal recessive conditions. These are conditions in which an affected person has changes in both copies of a gene that is located on one of the autosomes (chromosomes 1–22, i.e. not the sex chromosomes). A person who has one normal and one faulty copy of such a gene suffers no ill effects; such a person is a carrier for the condition. It seemed very likely that if Watson carried ten that were already known to genetic medicine, he probably carried at least a few others that hadn't yet been reported. The problem with this finding was that, by studying what happens when first cousins and other relatives have children together, the longstanding scientific best guess was that each of us is a carrier for perhaps one or two different recessive conditions. Studies in fish came to a pretty similar conclusion,[4] interestingly. So — why did Watson carry maybe ten times as many recessives as expected? To say the authors of the paper attempted an explanation is perhaps a bit of a stretch: they essentially said, 'he just happens to have a lot of these … and maybe other people do too'.

The alternate — correct — explanation would become apparent over the next few years. Watson carried ten different genetic variants that had been reported in scientific publications as disease-causing. Today, only *one* of those is still thought to

---

4   Fish were caught in the wild, then allowed to breed, with brother–sister matings set up in order to see what would happen. This sort of thing is frowned upon in human genetics.

be disease-causing.[5] The others are harmless variants that were mistakenly reported as troublemakers.

How was it possible for this to happen? The root cause of the problem is the enormous amount of variation in the human genome. If we were to compare your genome and mine, there would be three *million* places at which we differed from each other, and similarly each of us has millions of differences from the 'reference' genome. There is no one 'normal' human genome — if there are 7.7 billion people in the world today, there are perhaps 7.65 billion different human genomes (accounting for identical twins, who share their genomes). The 'reference' genome is a standard of sorts, but differences from that reference are not necessarily abnormal — in fact, almost all of the variations each of us has is harmless. Some are mildly helpful to us, some mildly harmful, and only a tiny proportion have the potential to cause a recognised genetic condition. A lot of that variation sits in between the genes, or is in a gene, but not in a place that affects the protein code for that gene. When we sequence a person's exome, we tend to find about 40,000 variants that result in a difference in the parts of genes that code for protein. Some of those are very common, some are rare, and some are apparently unique. Even today, if we sequenced your genome, it is virtually certain that we would find numerous genetic variants that had never been seen before. The only exception to that would be if members of your family, particularly your parents, had already had sequencing done.

Suppose you do exome sequencing in someone who has a condition you think likely to be genetic, and caused by a change in a single gene. The starting point is that you have to sift through 40,000 protein-changing variants to find the one or two that you are looking for. You are looking for needles in a big stack of needles.

James Watson's genome was sequenced before many people

---

5   'Disease-causing' only when there's also a mutation affecting the other
     copy of the gene. If someone is a carrier of such a variant, and their other
     copy of the gene is normal, they will suffer no ill effects.

had had exome sequencing done, and well before large databases of exome and genome data became available. Watson's ten recessives had all been found in people with genetic conditions, and reported in papers published over the preceding decade or so. For various reasons, every single one of those reports was, not to put too fine a point on it, wrong. For example, Watson was found to have a change in a gene linked to severe eye disease, called *RPGRIP1*. He had one copy of the gene with the usual amino acid found at position 547, alanine, and one copy with a different amino acid, serine.

In 2003, researchers from Pakistan reported a family in which eight members of a large family with a degenerative eye disease all had two copies of the same change found in Watson; it also cropped up in two smaller families. At the time, the standard way to check if something you've found is just a normal variation was to check 100 people from the same population ('population controls') to see if they also had the change. Sequence 100 people and you get 200 copies of the gene, so you'd think that if something is common and harmless in that group of people, you'd have a good chance of picking it up. To save money, the researchers used a cheap screening test rather than reading the sequence of the gene — and in hindsight, that test must have failed, because they didn't find the variant in *any* of the controls.

Apart from getting the screening test wrong, you can't blame the Pakistani group for thinking they had found the answer in their patients. Alanine and serine have some different chemical properties, although they are far from being the most dissimilar pair of amino acids. Finding the same genetic change in 12 different people (across the three families) with the same condition would usually be very strong evidence of a link to that condition.

By 2005, a Dutch group had already reported that the variant was way too common to be a cause of a rare eye condition, but this information must have been missed by the team who sequenced Watson's genome. Thanks to Daniel MacArthur and his team, we now know that the variant is common in much of the world; nearly half of the people from a European background in the gnomAD database have one or two copies of the variant, and there are nearly 7,000 people in the database (out of 140,000 total from all backgrounds) who have two copies of the variant — i.e. both of their copies of the gene are this version of it. It's simply impossible for this variant to be the cause of a rare condition; and it's not even slightly surprising that you might find it if you sequence the genome of someone from a European background, like Watson.

Over the past decade, it has become uncomfortably obvious that it is very easy to get it wrong in genetics. It's tempting to criticise the Pakistani group whose screening test went wrong (the variant is nearly as common in South Asians as it is in Europeans, so, if their test had worked, they would surely have found it in some of their 100 controls), but the problem has turned out to be widespread across the genetics literature. Population data alone aren't the whole solution, unfortunately, because there is variation that is harmless but rare, as well as variation that is harmless and common.

The field of cardiac genetics has been particularly hit by the issue of genetic variants that are wrongly classified as disease-causing. In 2012 and 2013, a Danish group, led by Morten Olesen, trawled through the medical literature about inherited diseases of heart muscle and of heart rhythm, and compared variants that had been published as disease-causing against the very first public exome database, the Exome Variant Server. This had information from just 6,500 exomes but was a treasure trove of information when it was first released. Olesen's team found that the cardiac genetics literature was littered with mistakes, with many of the 'disease-causing' changes being far too common in the population. They calculated that if all of these were really disease-causing, one in four people would have a heart-muscle condition, hypertrophic cardiomyopathy; one in six would have another, dilated cardiomyopathy; and one in 30 would have long QT syndrome, a problem with the heart's electrical rhythm. The implication was clear — a *lot* of the variants that had been described as harmful were, in reality, harmless.

Perhaps even worse, it became apparent that not only are some *variants* wrongly reported as disease-causing, there are plenty of *genes* that have been wrongly associated with conditions. Sometimes, this comes in the form of a single report that is never replicated. This is (mostly) fairly harmless. Sometimes, however, genes find their way onto lists and into tests for conditions despite very limited evidence. Again, cardiac genetics is a problem area. For example, the genes *CACNB2* and *KCNQ1* are both commonly included in panels of genes for testing people with hypertrophic cardiomyopathy, despite there being only a tenuous link between these genes and this condition. This runs the considerable risk that people who are tested will be told that they have a variant in one of these genes, and that it is the cause of their heart condition. All sorts of bad information can flow from this — particularly, family members who do not have heart problems being tested and wrongly reassured, or wrongly told they are at risk.

The discovery that we had, as a field, been getting it wrong so often led to a worldwide swing towards caution in interpreting genetic data. While this is mostly appropriate, it carries its own risks. There are all sorts of potential consequences when we make mistakes, and it cuts both ways: there are harms from wrongly reporting that a variant is harmful, and harms from wrongly reporting that a variant is harmless. If we tell someone their child has a condition that they actually don't have, it might lead to mistakes in treatment. It might mean we give them wrong information about the chance another child might be affected; they could have a prenatal diagnostic test that leads to termination of a normal pregnancy, or a failure to identify an affected fetus. If we wrongly fail to report a variant that is in fact disease-causing, on the other hand, that might also mean someone misses out on treatment that might help them. It might mean that someone who could have been reassured that they had a very low chance of having another child with a severe condition misses out on that reassurance. If you're worried about this happening and it affects your thinking about having more children, we describe you as having 'lost reproductive confidence' — and it might mean you also lose the chance to have another, healthy child.

So we have to be Goldilocks geneticists: we can't be too hot when we make a call on whether a variant is disease-associated or not, and we can't be too cold. The genetic porridge must be just right.

Getting it right can sometimes be really *hard*. It's easy if we have population evidence that says, 'This thing is just too common to be harmful.' It's easy if we see something that has been seen dozens of times before in affected individuals and never in the general population. It's most of the in-between that presents a challenge.

You might think that computers are the answer. If so, you wouldn't be the first. There is a long tradition of someone looking at the problem and thinking, 'I know! I'll write a computer program that can tell nice from nasty.' There have been a few different approaches to this, mainly directed at the scenario in which one amino acid is switched for another (generally, if the change is to 'stop', it's not as hard to work out what it means). Some of the programs look at the chemical changes. Some look at conservation through the course of evolution. By now, there are a LOT of organisms that have had their genomes sequenced. If you're interested in a change from — say — alanine to serine, you could get your shiny new program to have a look at that position in the equivalent protein in creatures that are less and less similar to humans, or at the same position in similar proteins (in humans and other animals).

If you'd done that for Watson's *RPGRIP1* variant, the news would have been equivocal: apes and monkeys all have an alanine at that spot, as do most rodents, and camels, cows, and killer whales; elephants and bats have an alanine, and so do aardvarks and armadillos. But squirrels and Cape golden moles have something different in that location; so do budgies and ducks. The star-nosed mole even has a serine there! It's one of the things it has in common with James Watson, along with being a warm-blooded, hairy, four-limbed animal. Not many moles have Nobel prizes, though. Overall, the evolutionary evidence is not especially strong evidence for this particular change being damaging to the protein (even if we didn't have the population information).

Sometimes, we see amazing conservation. For example, in a child with severe epilepsy, we found a change in a protein in which the particular amino acid that was altered (proline) was the same in every species that had been sequenced, back to oysters and amoebas. If nature thinks it needs a proline right *here*, and can't abide substitution in all the hundreds of millions of years since our paths diverged from amoebas and oysters … there's a good chance that amino acid really, really needs to be a proline.

Anyway, back to designing your computer program. You don't have to restrict yourself to chemistry or conservation; you could combine both. Or you could gather up a bunch of scores from other peoples' programs and combine them to help make your new score.[6] Now, calibrate your shiny new program on a big set of variants that you already know are harmful or harmless, check it against another set, give it a clever name, and publish a paper, explaining —

Explaining that, after all this effort, you've made something *just a little* better than what was there before. Of course, you won't frame it that way, but that's the best you can hope for, it seems. And just a little better really amounts to not much good at all. Pick any of the more than 20 such programs that are already out there, and you'll find that *everybody* has hundreds, if not thousands, of variants that the program thinks are likely

---

6   Or — why not? — you could compare the amino acid change with the hypothetical common ancestor between humans and chimps. I'm not making this up; this is part of the basis of one of the most successful programs, CADD.

to cause problems. They aren't too bad at picking the harmless variants, but considering that the starting point for any one change you find in a person's genome is that it's likely to be innocuous, that's not a great achievement.

The reason why all these efforts to write a program that can do the job have failed comes down to the nature of the task. It *looks* like what you need to do is to put the things you're classifying into one of two bins: an enormous industrial container full of benign and only mildly harmful variation, and a golden eggcup that contains the one or two variants that were the reason we did the test in the first place. You have 39,999 harmless apples and one orange. The problem is that your starting point in this apples-from-oranges task is not really a pile of 40,000 pieces of fruit. There are all sorts of different ways that changing an amino acid can cause problems.

Perhaps the chemical properties of the two different amino acids are so different that it causes the protein not to fold properly. Or perhaps a protein forms but is unstable and doesn't last long enough to be useful. Or the protein may form, but extra modifications that are needed — like tacking on sugars — can't happen. Or there is a perfect protein, but it can't get to the place inside the cell where it is needed. Or the amino acid change isn't the problem at all: the DNA change messes up the splicing process, causing a completely different type of issue. There are other possibilities, but you get the point. The computer programs are valiantly trying to sort fruit, but they are being presented with a mixture of fruit, roundish rocks, tennis balls, and sea urchins. It's no wonder they don't do a great job.

So, we have population data — powerful, but limited.[7] We have prediction software — slightly better than nothing. We have information that's published in the medical literature — that we know is riddled with errors. Not looking too promising so far, is it?

---

7   One of the limitations is that many populations aren't represented in the databases, so we don't know much about their normal variation. Arabs, for instance, and Pacific Islanders, and Indigenous Australians.

Fortunately, there are a number of other pieces of information that are useful *some* of the time. One of the most powerful of these is the information we get from the doctor who saw the patient. At the most basic level, if you are doing a test to find out why someone has severe epilepsy, and you find a variant in a gene that has only ever been linked to a skin condition, it's not very likely that you have your answer. Other information includes things such as how a variant tracks in a family, whether it affects a part of the protein known to be critically important, and (sometimes) tests that can directly measure the function of the changed version of the protein.

Put all of the available information together, and you may well be able to come up with a reasonable answer.[8] Classifying genetic variants is one of the most challenging and interesting parts of my job, although the pressure to stay in the Goldilocks zone can be stressful when it's a close call. Like most labs around the world, we put the variants we assess into one of five categories. Benign (class 1) is a variant that we are very confident is harmless, often because it's too common to be anything else (like Watson's *RPGRIP1* variant). Likely Benign (class 2) is a variant for which there's good evidence it's harmless, but not quite enough to put it in the Benign group. Pathogenic (class 5) means we're close to certain that the variant can cause problems. Likely Pathogenic (class 4) means that there's strong enough evidence that a doctor can use the information for making medical decisions, but there's not quite enough evidence to make it class 5. For both Likely Benign and Likely Pathogenic variants, there's an appreciable chance (which notionally can be as high as 10 per cent) that the truth lies in the other direction.

In the middle are the Variants of Uncertain Significance (class 3). This means what it says: we're not certain if this is a problem

---

8   There are various systems out there to help make these assessments. The most popular and widely used guidelines were published in 2015 by the American College of Medical Genetics and Genomics; they aren't perfect, but they are pretty good, and they have the advantage that everybody in the field knows about them, even those who don't personally use them.

or not. This classification has been described as genetic limbo, in the sense of the place you get stuck in rather than the bar you try to wiggle under. If there's not quite enough evidence to call a variant Likely Pathogenic or Likely Benign, or if there are pieces of evidence that point in opposite directions, then what you have is a VUS. The most important — and often difficult — decisions we make relate to variants that skate on the border between VUS and Likely Pathogenic. Make the wrong call, in either direction, and people may suffer because of it. These are the unknowns that keep me up at night. Have I wrongly called a variant Likely Pathogenic, leading a patient and her doctor down the wrong path? Have I wrongly classified one as a Variant of Uncertain Significance, denying a patient and her doctor options that might have helped?

And if that's not difficult enough, consider that everyone has not one but *two* different genomes.

6

Power!

The ships hung in the sky in much the same way that
bricks don't.

Douglas Adams

If, by chance, you are one of those who live in fear of an alien
invasion, you are way, way behind the times. The invasion
happened a very long time ago, and the aliens are already here.
They are living not just among us, but *inside* us.

The story of life on Earth — which is your story, and mine
— is very, very old. We don't know exactly when life began, but
one estimate is 3.8 *billion* years ago. Imagine that the ultimate
genealogist has volunteered to trace your family tree. Perhaps
you already know a bit about your family: you know your
parents, and their parents. You have some information about
your grandparents and maybe your great-grandparents, an old
photograph of stiffly-dressed, unsmiling people from a hundred
years ago. If you're a keen genealogist, you may have a family
tree reaching back past that — for a few hundred years, or even
further. But at some point, you're going to hit a dead end. The
earliest person whose name we know lived in Mesopotamia a
little over 5,000 years ago (his name was Kushim). Even if you
assume a generation every 20 years, that's only 250 generations
of ancestors ... and you can't ever trace your line back even

as far as Kushim, because we don't know anything about his family.

Our ultimate genealogist is going to take you back a *lot* further. Five thousand years of written history spans only the last 2 per cent or so of the time that modern humans have been around. Before them, there were protohumans, who in turn traced their origins through the primate line, and back to early mammals. Imagine if you could line up a series of photographs of your ancestors in one continuous line — all the women, perhaps — along a long, long wall. As you walked along that wall, starting with your mother and her mother, and hers, you'd see nothing but humans, or creatures that looked much like humans, for a very long way. If there were three pictures every metre, you'd see nothing but human beings for two kilometres. Somewhere around there, a gradual change might happen. The people — and they would still be people, just not quite *Homo sapiens* — would become shorter, and furrier; eventually, many more kilometres down the corridor, you'd reach the first of your ancestors to walk on two legs. Back, through millions of years; the earliest mammals emerged about 200 million years ago. Keep going, for thousands of kilometres, and the family photos change again, to creatures that crawled on the banks of some warm sea, 400 million years ago — closer to the fishes from which they descended. But you can still trace the line, from mother to mother to mother. The journey back from the first fish to leave the sea to the first fish worthy of the name is just another hundred million years or so. Back and back; 900 million years into the past, and we're down to the first, primitive multi-celled creatures.

The alien invasion happened *a billion years* before that.

For quite a long time now, your walk along the wall of family portraits has been something of a dull affair. Single-celled organism after single-celled organism, all looking much the same, with only the slightest changes over periods of millions of years. At last, though, you start to notice something strange. The single-celled organisms aren't really single cells. Inside the tiny creature, there are others. In fact, those others were always there but had seemed just another part of the whole, unremarkable little blobs within each cell. Now it's clear that these objects are *different* from that single-celled creature's other components. If it were a video portrait rather than a still picture, you'd see that they move around freely within the cell; it's clear that, although comfortably at home, they are leading a life of their own. Perhaps they even look to you like parasites rather than part of the whole. They are very far from that, however.

And now something truly strange happens. Your family tree *splits*. Even though you had followed mother after mother after mother, choosing just one parent at each generation, and continued to do so until you reached, perhaps already, the origins of sex itself ... now there is a *double* row of portraits stretching out along the wall.

You've just walked past a pictorial history of one of the most important events in evolution, happening in reverse as you backtracked through your ancestry. What happened, nearly two billion years ago, was a joining of forces between two primitive organisms. One, the larger, was from a line that had already been building in complexity. Around the same time as this remarkable union, the nucleus formed. The larger partner was one of the first of the eukaryotes, the group of organisms that includes almost everything that isn't a bacterium. After perhaps a billion years of nothing much happening, an evolutionary cataclysm was underway. The other partner was a much smaller and simpler organism, but had a neat trick up its sleeve (not that it had sleeves, of course. Or arms).

The smaller creature's neat trick was that it was really good at harvesting energy from food. Its new host could do this, too, but much more slowly and less efficiently. So there was a win for both sides in the new arrangement. The smaller partner gained protection from others who would eat it, and perhaps a steadier supply of raw materials than it could forage on its own. The larger gained a burst of energy that would fuel the next two billion years of evolution.

Not that long after this event, in another line of cells from ours, the *same thing happened again*. This time, the new cell that took up residence had a different specialty: it could take water and carbon dioxide, and, using energy from the sun, manufacture food. The first plant had been born.

Now, we call the remnant of that small invading bacterium the mitochondrion (the second set of invaders, which make a plant a plant, are called chloroplasts). Mitochondria have changed a great deal since they first took up residence in our cells, but, in some ways, they are still independent creatures, with their own DNA, their own slightly different metabolism, and their own reproductive cycle. We are utterly dependent on our mitochondria — we need them to survive and thrive. When things go wrong for them, they go wrong for us — often very badly indeed.

There is something fundamentally odd about this arrangement. It's as if you went swimming one day, and a small but particularly aggressive eel latched onto your belly and managed to *eat its way inside you* ... and then took up residence. Then, instead of killing you, the eel became one of your most important organs. In this imaginary world, the eel regards this as an exceptionally good deal. Did I mention that it was pregnant at the time, and now it has a nice safe place to raise its babies? But everything's okay, because *you are also happy* about the situation.

I don't know about you, but I can never quite get used to this idea.

We should not imagine, by the way, that this was a one-off event, a happy chance that worked out perfectly the first and only time it occurred. Countless small blobs of life consumed even more numerous, smaller blobs of life, over and over for hundreds of millions of years. During that time, there must have been many occasions when a partnership *almost* formed, when for perhaps as many as a few generations of cells something close to stability was generated — only for the big blob to lose patience and consume the smaller, or for the small to grow too fast and consume its host from the inside. But in the end, the partnership formed and was a roaring success. Look around you: everything you can see that is not rock, sand, or water is the result of that partnership. Every bush, every tree, every coral reef, every house, road, ship, and plastic bag ... all exist only because of this union between your ancient ancestors, tiny dots of life that were just that bit more successful working together than apart.

Over the millions of years, our mitochondria have settled down and really made themselves at home. Gradually, they handed some key tasks over to the nucleus of the cell, and long, long ago lost the ability to exist as free-living organisms. They have kept one reminder of their free-living past: their own small genome. In humans, this is tiny: only 16,569 bases long and containing just 37 genes. Compare this with the genomes of modern bacteria: free-living bacteria seem to need at least about 1,500 genes, spanning about 1.5 million bases of DNA, and some have genomes five times that size. There are some bacteria with smaller genomes, but they depend on other organisms for their existence. Take *Mycoplasma genitalium*,[1] which has only about 470 genes. It can't break food down into components it can use, so it exists only as a parasite inside other cells. There are quite a lot of parasites like this — perhaps they are the near-misses of evolution, the scenario in which the 'one organism living inside another' arrangement worked out better for one partner (the parasite) than the other (the host).

So, the mitochondrial genome is a ghost of what must have been its former self. Of its 37 genes, only 13 code for proteins — the rest produce machinery needed to make those proteins, running a slightly different biochemistry from the nuclear genome in order to achieve this. It's possible that, eventually, all of those 13 genes will transfer across to the nuclear genome,

---

1   As its name suggests, it likes to live in human genitals. You definitely don't
     want it living in yours.

and the mitochondrial genome will lose its function.[2] In the meantime, mitochondria carry on behaving a bit like they are a separate organism. Each cell contains hundreds to thousands of mitochondria. Each mitochondrion contains multiple copies of its genome,[3] which exists in the form of a ring of DNA — like the genomes of bacteria, and quite unlike the chromosomes in the nucleus. The mitochondria live their own little lives, semi-independently from the rest of the cell. They divide into two just like bacteria; they grow and age and die at their own pace. When cells divide, the new daughter cells share the mitochondria from the original cell between them. In order for dividing mito-chondria to have their own copies of the mitochondrial genome, that, too, has to be copied, and there is a special set of cellular machinery for that. Like any DNA that is being copied, there is a chance for mistakes — and so there can be mitochondrial mutations. They have some special properties, as we shall see.

We depend on our mitochondria: they do many tasks, but, most importantly, they are like generators making the energy our cells require. The digestive system, including the liver, breaks food down into its component parts. We can use some of those parts for energy: sugar, fat, and (at a pinch) protein can be burned for fuel. These raw materials get passed to the mitochondria for conversion into a form of instant chemical energy the cell can burn to do whatever it needs to do. This means the cells that use the most energy — brain, muscle, heart, and so on — are the most vulnerable if something goes wrong

---

2   The degeneration of both the Y chromosome and the mitochondrial genome has been ascribed to Muller's ratchet. The ratchet is a gradual accumulation of harmful mutations in situations where chromosomes can't trade information. During formation of eggs and sperm, chromosomes 1–22, and the X chromosomes in egg production, undergo a process called recombination, in which material is exchanged between the two versions of the chromosome. The Y can only do this in a very limited way, and mitochondrial DNA can't do it at all, leaving them vulnerable to decay over the millennia.

3   There's a special case, though: in the egg, each of the 200,000 or so mitochondria contains just one copy of the mitochondrial genome.

with the mitochondria. By contrast, cells that don't need a lot of power — skin cells and fat cells, for instance — can cope relatively well when their mitochondria fail.

Something was wrong with Felicity's mitochondria, and had been for her whole life. It was only now, though, in her late 30s, that the first signs of this were beginning to show themselves. The progression was so subtle, so insidious that Felicity didn't even notice, until one day her husband pointed out that her eyelids had begun to droop. She saw an eye specialist, who realised that the problem wasn't just in her eyelids. Her eyes were not moving as well as they should, because slowly her eye muscles were getting weaker. Eventually, she would not be able to move her eyes at all, and would have to turn her head to look to the side. Otherwise, though, she was completely well.

Ahmed had always been a clumsy child, and had struggled to keep up with his friends in the playground. From the time he was about ten, his mother started to worry that he was increasingly unsteady on his feet, and he had begun to walk on his toes. He wasn't doing well at school — everything seemed more difficult for him than it had when he was younger. His mother took him to a paediatrician, who found that Ahmed had some muscle weakness, and that he was walking on his toes because of tightness in his Achilles tendon, which made it hard for him to get his feet flat to the ground. Ahmed, too, had drooping eyelids and could not move his eyes normally.

Jacob did not live to see his first birthday. Very early in life, he was found to have anaemia — his bone marrow could not make enough red blood cells to meet his body's needs. It was so severe that he needed regular blood transfusions. His pancreas did not work as it should, so that he could not absorb food well. There was a high level of lactic acid in his blood, and his liver was sick from the day he was born, until it failed him completely late in his first year.

Jacob, Ahmed, and Felicity, although so different, all had exactly the same underlying problem: some of their mitochondria were missing a wedge from their circular DNA. For reasons we don't understand, if you are missing a large part of your mitochondrial DNA, you're quite likely to develop one of these three related conditions — chronic progressive external ophthalmoplegia (CPEO),[4] like Felicity; Kearns-Sayre syndrome, like Ahmed; or Pearson syndrome, like Jacob. Had Jacob survived infancy, he, too, would almost certainly have eventually developed similar eye problems to those that Ahmed and Felicity had. Although these three conditions have different names, really it's all one condition with differing degrees of severity, from CPEO at the mildest end to Pearson syndrome at the most severe.

Genetics is full of situations in which changes in DNA that you might expect to have wide-ranging effects instead produce oddly specific manifestations, and this is a typical example. Having a change in just one of the 16,569 bits of DNA that make up the mitochondrial genome can have a devastating effect on multiple organs, sometimes causing death in the first days of life. So why on earth should removing as much as three-quarters of the mitochondrial DNA lead to someone having only an eye condition, and one that takes decades before it becomes noticeable? And why is it that Jacob died in infancy, Ahmed had significant and varied medical problems in childhood, and Felicity was completely healthy apart from her eye condition?

We only partly know the answer to these questions. The bit that we understand best has to do with the fact that the mitochondrial genomes inside a cell are independent from each other. This means that it is quite possible for a single cell to have more than one version of the mitochondrial genome. That could

---

4   This sounds daunting, but it's just jargon. Chronic means present over a long time. Progressive means getting worse over time. External means outside. Opthalmo refers to the eye. And plegia means weakness, as in paraplegia. Put that together and you have a long-term condition that gets worse over time and affects the muscles on the outside of the eyeball (the muscles that move the eye).

mean harmless variation — or it could mean that there are some copies that are normal, while others have something wrong with them, such as a missing piece. There's a term for this type of mixture: heteroplasmy. It's easy to imagine that, if most of the copies of the mitochondrial genome are normal, that might lead to milder problems than if almost all of the mitochondrial DNA is abnormal.

At least in part, that explains the difference between Felicity and Jacob: if we could go inside each of their cells and count normal and abnormal copies, Felicity would very likely have many normal copies, but Jacob would mainly have abnormal copies. The fewer normal copies that are present, the worse the function of the mitochondria, with correspondingly more severe health problems that start earlier in life.

That leads to another question. How was it possible for Felicity to be perfectly fine for decades, despite her cells containing at least some faulty mitochondria? The answer is that as we age, our mitochondria accumulate damage, including deletions like those seen in CPEO. Felicity started life with a certain percentage of damaged copies of the mitochondrial genome. She had enough normal, functioning copies for the cells of her eye muscles to work just fine. Over the years, damage accumulated and those cells passed some threshold of damage, past which they could no longer cope. Slowly, her eyelids began to droop. Poor Jacob was born with his cells already past the redline — most of his mitochondrial DNA was abnormal.

Different types of cell are better at tolerating this problem than others, which explains the different pattern of symptoms in our three patients. In Jacob, many different tissues were struggling from the start, because their cells were beyond the level of faulty mitochondrial DNA at which they could cope, and there was no way back. Ahmed was in between the two extremes. The mild clumsiness his mother had noticed early on was a sign that there were some nerve cells that were struggling a bit. Even from birth, he must have had many cells that were only just keeping up with the demand for energy. Just a few years later, a little extra damage had happened and his symptoms began.

At this point you might be wondering about your own mitochondria. Are they also accumulating damage? You bet they are. In a 2006 study, researchers from the University of Wisconsin looked at muscle samples taken from people aged from 49 to 93 with no known mitochondrial condition. They found muscle fibres with faulty mitochondria in *all* of the people they studied, and the older the person, the worse the problem. In people in their late 40s and early 50s, about 6 per cent of muscle fibres showed abnormalities; for those in their 90s it was 30 per cent. They also found a steady increase in the level of deletions similar to those seen in people like Felicity, Ahmed, and Jacob (but at much lower levels). The loss of muscle strength we all experience as we age is partly down to this gradual decay of mitochondria. You can't expect a faulty powerhouse to supply the energy a muscle needs.

If individuals accumulate mitochondrial damage, what about the species as a whole? How is it possible for generation after generation of people to be born and live their lives, mostly not having any sign of mitochondrial disease, if their mitochondrial DNA is constantly decaying? Why didn't we all go extinct millions of years ago?

Strangely, the answer lies in the neck of a bottle. In this case, it's a figurative bottle: the mitochondrial bottleneck is an event, rather than a physical thing. It's an event that happens very early in a woman's life, before she is even born, and it has to do with the number of mitochondria in the egg. A typical cell has thousands of copies of the mitochondrial genome, but a human egg cell has about 200,000 copies. After the egg is fertilised, the cells divide faster than the mitochondria can keep up, so that the number per cell drops quickly towards a more 'normal' level. Among the many different types of cell in the growing (female) embryo, there is one type that will eventually become the future eggs; the next generation is being plotted almost from the start.

This process involves many cell divisions. Somewhere along

the way, the number of mitochondria drops dramatically and then expands again: this is the bottleneck. We're not sure quite how low the number goes — how tight the bottleneck is — or even if it is a single bottleneck, or rather a process of widening and narrowing, a sort of wavy bottleneck. Regardless, the idea is that if you have a small number of mitochondria — maybe a couple of hundred — which are then multiplied to become 100,000, anything dangerous that lurks in the mitochondrial DNA in those 200 will also be multiplied. That might seem a bad idea, but it means that a slow accumulation of damage over generations can't happen.

The reason for this is that if any damaged DNA slips through the bottleneck, it will be amplified up and is likely to become a significant proportion of the resulting egg's complement of mitochondrial DNA. An egg with a high burden of damage is very likely to be prevented from passqing that damaged DNA to the woman's descendants. In the ideal world, this would be by making the egg non-viable — incapable of being fertilised, or of dividing and becoming an embryo. Perhaps that is what usually happens. But from the point of view of survival of the species, it doesn't make a great deal of difference if an occasional baby is born but never grows old enough to have children of her own. As long as that heritage of damaged DNA is not passed on, the species is saved from its effects and can continue. From this perspective, the existence of mitochondrial disease is the price we all pay for continuing to exist.

It is a harsh price for those who have to pay it personally. We don't know for certain why, but the kind of mitochondrial deletion that causes the problems we saw in Felicity, Ahmed, and Jacob is almost never passed on from mother to child. It seems likely that the bottleneck is very efficient at removing these deletions. This is not true for other types of mitochondrial mutation, however. A change in a single base of mitochondrial DNA can cause problems every bit as severe as those seen in Jacob, with the same variability due to different mixes of healthy and mutant mitochondrial DNA. A lot of the time, this seems to happen as a one-off: a particular line of cells gets unlucky and produces a single egg with a high load of abnormal mitochondrial DNA, and it never happens again in that family. Sometimes, though, we see multiple siblings affected, and even passage through several generations.

As you'd expect, this is more likely to happen if the problem is something that allows an affected person to have children of their own. One such condition is called Leber hereditary optic neuropathy (LHON). People with LHON are generally well and healthy, until, one day in their teens or 20s, they notice their vision is cloudy in one eye. Quickly, the symptoms worsen, until, at best, the sight on that side is barely enough to be able to count fingers held up before the eye. Then, a couple of months later, the same thing happens to the other eye.[5] Most affected people never have any improvement in their vision; most become legally blind. Occasionally, there can be neurological problems that go along with the eye condition, but, usually, vision loss is the only thing that happens.

There are some mysteries about LHON. One is that many people, despite having no normal copies of the mitochondrial genome, go their whole lives without losing their sight. Another is that men are far more likely to lose their vision: about half of men but only about 10 per cent of women with an LHON variant have vision loss. And lastly, when we can track the change through the family, we often see that the proportion of abnormal mitochondrial DNA — the 'mutant load' — rapidly increases to 100 per cent, usually within a generation or two. For this condition, the bottleneck seems to be eliminating normal rather than abnormal mitochondrial DNA. We have no idea why.

All of this makes discussions with families about the implications of the condition very different for LHON than for other genetic conditions. A woman who has LHON has a *one hundred per cent* chance of passing this on to each of her children

---

5   About a quarter of the time, both eyes go at once.

(an affected man has effectively zero chance of passing it on, because our mitochondria are inherited only from our mothers[6]). This means that all of her children will inherit the genetic change, but only half the boys and 10 per cent of the girls will lose their vision because of it. We can't tell in advance which ones, we can't predict exactly when they will have problems, and there is nothing that we know that can prevent it.[7]

For other mitochondrial mutations, it can be a different story.

By the time we made the diagnosis of Leigh disease in Joseph, his younger sister, Kylie, had already been born. In his first year, their mother, Pauline, had taken Joseph to see several doctors, who had reassured her that his slow weight gain and low muscle tone were nothing to be too concerned about. He was gaining skills normally, and otherwise seemed fine. Gradually, though, things got worse. Around the time of his first birthday, Joseph had a bout of gastroenteritis, with vomiting and diarrhoea that lasted for a week. He had finally started crawling just a few weeks before, and now he stopped. He would never crawl again. Over the next few weeks, he developed some unusual movements of his hands and his breathing took on a strange pattern, with bouts of hyperventilation. Heavily pregnant, Pauline took Joseph to a paediatrician, who was deeply worried by Pauline's story, and by what he found when he examined the child. Referral to a neurologist followed quickly; she ordered scans and blood tests. Finally, a biopsy of muscle and liver was done, to measure the function of the mitochondria. This test result was abnormal, telling us that the mitochondria were not functioning as they should.

By now, Kylie was two months old. She was there in the room

---

6   This is a general rule — as for so many things, there have been a couple of exceptions reported, in which paternal mitochondrial DNA somehow ran the gauntlet of conception and survived to contribute to the resulting child's genetic makeup.

7   There is some general advice for at-risk people, such as not smoking and avoiding excessive alcohol, that might help a bit.

when we discussed the results with Pauline and her husband, Mark. We knew that Joseph had a mitochondrial condition, but didn't yet know the genetic basis for this. Almost any type of inheritance was possible, since the mitochondria rely so heavily on input from the nuclear genome. But the testing showed that nearly all of his mitochondria carried a genetic change, a single alteration that changed a T to a G. This was a change that had been seen in many other people with Leigh disease, and confirmed everything that the neurologist had feared. The diagnosis of Leigh disease meant that Joseph had a degenerative condition that would relentlessly worsen, taking away the skills he had learned and stopping him from gaining new ones, affecting his breathing and his ability to swallow. Like many children affected by this terrible condition, Joseph would not live to see his third birthday.

Often, this type of problem is a one-off in a family. Often — but not this time. Over the next few months, Kylie, too, showed signs that she might be affected. We went straight to the DNA test this time. This showed that almost every copy of Kylie's mitochondrial genome had a G at position 8,993 instead of a T. The news could not have been worse.

We tested Pauline next, and found that 35 per cent of her mitochondrial DNA had the change. An adult neurologist examined her and found no trace of any ill effect on her own health from this.

It took nearly a year of grieving and learning to live with the truth that her two children were dying of a condition we were powerless to treat before Pauline and Mark could start to think about the future. Finally, though, they came back to clinic and asked what they could do to have a healthy child.

As we'll see in chapter 7, there's more than one way to make a baby. We talked about using donor eggs; they considered this, but asked about having IVF and testing the mitochondria in the embryos, then implanting only an embryo that was predicted to be healthy. This process is called pre-implantation genetic testing (PGT). PGT is commonly used in situations where there is a known genetic condition in a family — for example, if both parents are carriers for a recessive condition — and can also be used for checking embryos' chromosomes. At the time of these events, PGT for mitochondrial conditions had only been technically possible for a few years. Early on, there had been concern that the cells of an embryo that had only gone through a few cell divisions might have varying levels of abnormal DNA, so that the few cells that were biopsied for the PGT procedure may not have given an accurate picture of the whole embryo. Fortunately, this turned out not to be the case, and, by the early 2000s, there were even statistics for this specific variant that gave a guide to what might happen for any given mutation load, allowing us to turn a percentage into useful information. From around 60 per cent load, the likelihood of severe problems affecting the child rises steeply.

At Pauline's first cycle of IVF, *all four* embryos had a load of more than 95 per cent. We thought it might never be possible for Pauline to have a healthy child of her own, but she tried again. This time there were six embryos. Two had a load of more than 95 per cent; three had similar loads to Pauline herself — but one had less than 5 per cent. A successful pregnancy followed, with the birth — against considerable odds — of a healthy girl.

It's obvious, though, that things could easily have gone differently. Pauline might never have had an embryo with a low risk of disease; there have been others who found themselves in that boat. In the early 2000s, when this story happened (Pauline's daughter is a teenager now), if that had been the case, there would have been no option for Pauline to have a healthy child who was biologically hers. Today, a new option is becoming available for families like this. It's yet another way to make a baby.

The idea has been kicking around for at least 20 years. Every five years or so, it pops up in the press as a startling new idea, with much fulminating about the ethical issues it supposedly raises, but in the last few years it has become a practical option. It's a simple enough concept: if a woman's eggs contain faulty mitochondria (let's call her 'woman A'), why not get healthy mitochondria from another woman ('woman B')? In practice, because our cells are not just bags of fluid but have a complex internal structure, it turns out to be much easier to do it the other way round and transfer the cell nucleus from woman A into an egg from woman B — removing the nucleus from woman B's egg first, of course — and then fertilising the resulting egg. Alternatively, the egg from woman A can be fertilised and then the pronucleus — the combined egg + sperm nucleus — can be transferred into the donor egg from woman B. Either way, what you wind up with is nuclear DNA from the couple who are trying to have a healthy baby, and mitochondria from another woman. A few abnormal mitochondria are likely to tag along with the nucleus, but, as long as they make up a tiny proportion of all the mitochondria in the resulting child, it shouldn't be a problem.

We know this can be done, because there have been animal experiments over a number of years, as well as increasingly ambitious experiments with human embryos. Reportedly, there was an attempt in China in 2003 in which a woman conceived twins. They were born prematurely and died, although their mitochondria were said to be normal. In 2016, a group from New York announced that they had performed the first successful procedure, in a woman from Jordan who gave birth to a healthy baby. The work was done in Mexico, not in New York — because all such procedures are banned in the United States, as they are in many countries. One of the few places mitochondrial transfer is permitted is the United Kingdom. The Human Fertilisation and Embryology Authority (an Orwellian name if ever there was one) conducted reviews and public consultation in 2011, 2013, 2014, and 2016 and finally, in December 2016, gave approval for 'cautious' use in 'specific circumstances where inheritance of the disease is likely to cause death or serious disease and where there are no acceptable alternatives'.

There are some real concerns about the use of the technology, and also some rather silly ones. The main real concern is the same as for genetic editing of an embryo's DNA: safety. On the face of it, the risks seem lower for mitochondrial transfer, because nothing is being altered — it is a transfer of something we know is functioning normally, from one cell to another (the term 'mitochondrial transfer' is a bit misleading because it's the nucleus that's transferred, but in concept that's what happens). Yet there are still ways that things might go wrong, with consequences for the resulting baby.

Less sensibly, there has been a lot of arm-waving about 'babies with three parents'. Some people are concerned that this medical procedure might be 'playing God', although it's hard to see that it's more God-like than any other form of artificial conception. Others worry that this could put us on a slippery slope to designer babies. In this case, the babies are 'designed' to have mitochondria that aren't faulty, making them essentially indistinguishable from almost everyone who wasn't designed. It's a bit hard to take that one seriously. And some suggest that a person born with genetic material from three parents might have a conflicted sense of self.

That last one is perhaps best answered by bringing our ultimate genealogist back. This time, we're not going to restrict the UG's activities to just the female line — we're going to look at *everyone*. How many ancestors do you think you have? Well, let's be conservative this time and assume a 30-year generation time. Say, three generations per century. Go back 100 years and we're only up to eight ancestors. By 200 years, it's 64. But as we go further into the past, the numbers ramp up quite quickly. At 500 years, we're up to 32,768 ancestors: a good-sized country town.

Why not have a guess at how many there are by the time you have gone back 1,000 years?

.

.

.

.

.

Okay, how did you go? It turns out the answer is 1,073,741,824. More than a *billion* ancestors. But since the entire population of the world only passed a billion people sometime in the early 19th century, and your ancestors are unlikely to have been evenly distributed across the world, it's apparent that there must be a LOT of people who crop up in your family tree more than once. Yes, many of your ancestors were undoubtedly related to one another. Moreover, you are related to pretty much everyone who hails from the same part of the world that you do (and to everyone else in the world as well, although the link may be as far back as tens of thousands of years ago).

So ... what possible difference could it make that you (sort of) have one extra relative in the most recent generation? It's of no consequence at all when we consider your truly enormous family tree. And it really is only 'sort of'; the tiny genetic contribution of the mitochondrial genome pales into insignificance beside the mighty nuclear genome.

People in Pauline's unfortunate situation are rare, so there are no floodgates to open. Nonetheless, there is undoubtedly a demand for the procedure, and it is likely that people will calm down about it once it's been going for a few years. It's easy to forget how controversial in-vitro fertilisation was when it was first introduced. There were protests at conferences and outside IVF clinics — including, oddly, by 'right to life' groups. Now, IVF is routine. There's a good chance you know someone who has used IVF in an attempt to conceive. Given the rarity of the problem that it treats, this may never be the case for mitochondrial transfer — but it is likely to similarly become routine, a non-issue.

You might reasonably think that since we know so much about the biology of mitochondrial diseases, we might have some treatments that work. Disappointingly, we mostly don't. There is some evidence that supplements with a substance called creatine — beloved of bodybuilders, with the helpful side effect that it's possible to buy the stuff at reasonable prices — are useful for people with muscle weakness due to mitochondrial disease. Many people still treat people with mitochondrial disease by giving them a 'cocktail' of vitamins, chosen as antioxidants and in the hope of giving a boost to parts of the mitochondria that are failing. Occasionally, you hear of someone who seems to respond very well to this — but the problem is that there is no way of knowing what would have happened had that person not been treated.

This point was brought home to me in a startling way in 2012, when I was contacted by Professor David Thorburn about a patient called Brandon who had been seen at Sydney Children's Hospital more than 20 years previously, before I worked there. David is a luminary in the field of mitochondrial disease, esteemed for many reasons but particularly because he never, ever gives up on the possibility of making a diagnosis. A neurologist, since retired, had sent David samples for testing in the early 1990s, and now, at last, there was a likely genetic diagnosis. Could I provide more clinical information, and see if I could contact the family to let them know the news?

I dug out the old notes from deep storage. Brandon had started having trouble breathing in the first days after he was born. The levels of lactic acid in his blood and in his spinal fluid were very high, which can be a sign of sick mitochondria. He was floppy, and although eventually he started breathing well enough that he no longer needed artificial support of his breathing, he did not suck or swallow, so he had to be fed by tube. The levels of lactic acid in his blood remained sky-high throughout the time he was in hospital. The second-last entry in our notes said that he was being transferred back to his local hospital to be close to his family. The very last entry in our notes concerned the results of muscle and liver biopsies, which showed that his mitochondria were not functioning properly, confirming the general nature of the problem (but not telling us the exact genetic cause). The story reminded me of other newborn babies I had seen with severe mitochondrial problems, and I was sure that Brandon could not have lived much longer.

After 20 years, it took a bit of detective work, but I finally managed to get hold of Brandon's mother. I wasn't sure how she would take this call out of the blue, and was worried about the grief I might be causing her, reminding her of this long-lost baby. I need not have been concerned: her disposition was sunny, and she seemed pleased to hear from me. Then ... she offered to give me her son's phone number! He had made a full recovery from his early problems, had grown up and gone to school, and was now working in the family business. Soon, to my amazement, I found myself talking to the man himself.

Soon after, I heard of another baby who had exactly the same genetic change as Brandon, with similar early problems — but who had indeed survived only a few short months. You may well ask how all of this was possible. I certainly did. Why did one baby die while another, seemingly just as sick in the first few weeks of life, go on to not just survive, but thrive?

It turned out that Brandon had inherited a faulty copy of a gene called *LYRM4* from both parents. The role of this gene has to do with the metabolism of sulphur in the body, which in turn is important for the function of the mitochondria. Newborn babies do not handle sulphur very well, but, over the first few months of life, the relevant systems mature rapidly. Make it past those first few months and it seems like you can cope pretty well despite a faulty *LYRM4* gene.[8]

If we had treated him from soon after birth with a mito-chondrial cocktail and seen stabilisation, then improvement of his condition, it would be awfully tempting to think that it was our treatment that had made the difference. Tempting, but completely wrong. This doesn't mean that mitochondrial cocktails

---

8   This particular cause of mitochondrial disease appears to be very rare, and
     recoveries like Brandon's are truly exceptional. Still, as David Thorburn
     and his team suggested in the paper reporting this discovery, it might
     be possible to treat babies in this rare situation by giving them specific
     sulphur-containing supplements. This doesn't seem to have been tested yet,
     which isn't surprising — only three people have ever been reported with
     problems due to *LYRM4* variants.

couldn't sometimes help people ... but mostly, it seems like they just don't work.

On the other hand, it seems that we are just about to enter a golden age of new, targeted treatments for genetic conditions. For some, this might even mean that most elusive of goals in medicine — a cure.

7

Dysmorphology Club

I don't want to belong to any club that would accept me
as one of its members.

GROUCHO MARX

The face is the mirror of the mind, and eyes without
speaking confess the secrets of the heart.

ST JEROME

When I first met her, 20 years ago, Dianne was a chubby, cheerful infant. She had been giving her parents and her doctors a good deal of concern, ever since she was born prematurely, a full ten weeks before her due date. Born weighing 2.2 kilograms, she was much heavier than you might expect — nearly a kilogram above average for such a premature baby. It's no bad thing to be bigger than expected, when you're born prematurely. But a whole kilogram over the odds was noteworthy, and her doctors took note. They also saw that she was unusually hairy for a newborn. More worryingly, she had a loud heart murmur. After a while, her heart began to fail, because it had not managed to adapt to life outside the womb.

You often hear the heart referred to as a pump. Actually, it is *two* pumps. One (the right side of the heart) collects blood from the body and sends it to the lungs to be scrubbed of carbon dioxide and enriched with oxygen. The second, larger pump (the left side of the heart) collects that revitalised blood from the lungs and sends it back to the body. Before a baby is born, however, things are different. The fetus gets its oxygen from the placenta, which also passes nutrients from the mother to the fetus, and cleans its blood — it's like lungs, liver, and kidneys rolled into one. The placenta doesn't get nearly the credit it deserves — calling it 'afterbirth' seems disrespectful, when you consider the remarkable and essential job it has been doing before birth.

While the placenta is doing its job, the lungs are breathing liquid, not air, so spending a lot of energy to pump the whole of the baby's blood through the lungs would be wasteful. Instead, there are two bypasses, so that most of the blood that has been charged up with oxygen and nutrients by the placenta can skip the lungs and go straight to the body, where it is most needed. One of these bypasses is inside the heart: a channel, called the foramen ovale, from the right atrium (the blood-collecting chamber) to the left. The other bypass is outside the heart, a short artery that links the aorta — the arterial freeway that emerges from the heart and carries blood to the body — to the pulmonary artery — the equivalent artery carrying blood to the lungs. This is the ductus arteriosus, and it's an important piece of plumbing.

Until birth, that is. At a baby's first breath, things shift inside her heart. The pressure on the left side of the heart goes up, slamming shut a flap of tissue that covers the foramen ovale. And the ductus arteriosus spasms shut. In most people, the foramen ovale soon seals closed, and the ductus does, too, eventually withering away to become a humble strip of gristle linking the two great arteries.

But if it doesn't ... there might be trouble. Blood pumped to the aorta whooshes back through the ductus to the lungs, meaning that the heart has to pump it all over again. If the ductus is wide open — as it was in Dianne — the extra work can just be too much for the heart to cope with. Sometimes, medications can persuade a reluctant ductus to close. In Dianne's case, this didn't work, so that, at just two weeks of age, her tiny chest had to be opened by a surgeon, who closed the duct.

For a while, things seemed to go well, but by nine months it seemed that Dianne was struggling to breathe; tests showed that the right side of her heart was having to work much harder than it usually should. On top of that, she was not meeting her milestones — she was late to sit, late to stand, and, in time, late to walk.

Dianne's doctors thought she probably had a syndrome — but which one? It was my job to figure it out, and I hadn't a clue.

To most people, if the word 'syndrome' means anything at all, it conjures up thoughts of Down syndrome, Tourette syndrome, or even Stockholm syndrome. The villain in *The Incredibles* named himself Syndrome. Applied to your child, it has a frightening sound. But to a doctor, 'syndrome' just means a collection of features that are seen together, and there are literally thousands of different syndromes, most of which are very rare. Some are mild,[1] some are severe, and many of them are variable: sometimes mild, sometimes severe, sometimes in-between. Even within the same family, we see great variability — one person may have almost no features of a syndrome while another has a lethal condition, both caused by exactly the same change in the same gene.

One of the main ways we diagnose genetic syndromes — until recently, almost the only way — was simply to look at our patients, trying to recognise distinctive patterns. This is a skill called dysmorphology — a terrible word, because its literal meaning is *the study of bad shapes*. Not something you want

---

1   Your idea of what constitutes a 'mild' condition may be quite different from mine. The distinction may seem arbitrary, but, as we will see in chapter 11, it can be quite important.

applied to yourself, and still less to your child. It might be better named 'pattern-ology' or perhaps 'protypology', if we wanted to stick with Greek roots.

The dysmorphologist — one who studies dysmorphology — looks for recognisable patterns, in the shape of a person's face and in their other features. The characteristics of hair, skin, nails, and teeth, the patterns of lines on palms and soles, and even fingerprints can all contribute to making a syndromal diagnosis, as can neurological and developmental problems, behaviour, and even patterns of sleep. Although professional dysmorphologists are almost all clinical geneticists by training, there is a real sense in which everyone who can see is a dysmorphologist, including you. If you have seen one or two people with Down syndrome, you are likely to recognise the condition in others, without needing a blood test to prove it. Even when you distinguish friends and family from acquaintances, you are employing the essential skills required of a dysmorphologist. But a clinical geneticist makes it her business to study rare facial patterns, particularly those that go along with other features such as intellectual disability. Sometimes, you see someone in clinic for the first time, and it's like running into an old friend on the street, or like that shock of recognition when you see a celebrity in the supermarket.

The spot diagnosis in genetics comes from that shock of recognition. We typically book patients in to see us for an hour, and often that is barely enough to learn all that we need to about our patient. We may need to consider a complex family history, learn the medical history of the person we are seeing, conduct a physical examination, and discuss our findings and planned investigations. So how do you handle the consultation if you've made a diagnosis the moment the patient walks into the room? Carefully! You can't say, 'Hi, it's nice to meet you. Your child has Noonan syndrome,' because it's not likely to go down well. As dysmorphologists go, I'm nothing special, but I've made the occasional spot diagnosis. It's surprisingly awkward conducting the usual preliminaries, when you already know exactly where the consultation is heading.

Even approached with care, people may not be that happy that you've diagnosed a syndrome in their child, or in themselves. Sometimes it's easy to understand why. A paediatrician once asked me to see a baby who had had low levels of sugar in her blood soon after birth. The problem had improved with treatment and now she was well. But the paediatrician considered the possibility that there might be an underlying problem with the way her body handled energy, and asked me to see her to think about whether any extra tests were needed.

There were some delays in making an appointment, so that Helena was nearly a year old when I first met her. I noticed she was very tall for her age, with a big head. She looked like her parents ... but she also had a broad forehead, widely spaced eyes, and a small jaw with a distinctive crease above her chin. In short, she looked exactly like other people with a condition called Weaver syndrome. I arranged a genetic test that confirmed the diagnosis.

Children with Weaver syndrome, as well as being large for their age and having a recognisable 'look', tend to be slow at walking and clumsy as children. It's common for them to have some degree of intellectual disability, but not inevitable — some affected people have normal intellect, and really just blend in to the normal population as they grow up. There's a long list of other health problems that have been seen in people with the diagnosis, including low blood sugar in the newborn period. Most worryingly among these issues, perhaps, is that people with Weaver syndrome seem to have a higher than usual risk of some types of childhood cancer.

Imagine your young daughter has had low blood sugar a few times in the first week of life, but now she seems fine. Your paediatrician sends you to a geneticist just to be on the safe side; you consider cancelling the appointment because you don't expect to learn anything important ... and the geneticist tells you this news. How would you feel about it? You could argue (and I would) that, on the whole, it's better to know. Perhaps the child has a heart condition that hadn't been diagnosed before, but now will be picked up, because we screen newly diagnosed children with Weaver syndrome for heart problems (Helena did turn out to have a minor heart condition, but fortunately not of a type that needed any treatment). Similarly, if you know that your child is at risk of problems with development and learning, you can monitor her progress, and step in early with therapy if needed. And just maybe, being alert to the chance that your child might develop cancer will let you identify symptoms sooner than you might have done, and so get earlier treatment. But it seems that the increase in cancer risk is quite small, so there's a good chance this part of the news isn't relevant at all.

There are some pretty clear downsides to knowing the diagnosis. You are far more likely to experience the harm of worrying that your child might get cancer than to experience any benefit from the information. Knowing she may have problems with learning is also likely to be worrying, and it's still possible that her development will be fine. What about stigma? How will others view your child, knowing she has a syndrome? How will it affect your relationship with her? Will you change your behaviour to become over-protective, perhaps? What will you tell your parents, your other relatives, your friends, your child's school?

None of this is straightforward. There are times when I sit with a family, knowing I am about to send all of this worry, uncertainty, and ambiguity into their lives, and I wonder if I'm doing the right thing. Might it not be better to leave these happy, loving parents at peace for a while longer? Isn't it better to wait until there's a problem before making a diagnosis?

Mostly, of course, it's not like that. People come to us looking for answers, and they rightly expect us to do our utmost to provide those answers — accepting the possibility of bad news as part of that. There is some comfort in knowing that, often, there are direct benefits to the child and family from having information, such as the possibility of finding out about an unrecognised health problem associated with the syndrome and treating it. And then there are those conditions that have a high chance of happening again in future children. Knowing this is helpful, because it gives couples choices they would not otherwise have (more on this later, particularly in chapter 11).

Even the best dysmorphologist can't expect to make a spot diagnosis for every one of the many, many different syndromes, so we have tools to help us. The earliest of these, naturally, were books. The most famous of the dysmorphology books is *Smith's Recognizable Patterns of Human Malformation*. First published in 1970, the book is a handbook of common and not-so-common syndromes. Grouped by major features, each syndrome has a page of description and a few photographs of people with the characteristic features of that condition. *Gorlin's Syndromes of the Head and Neck*, a massive tome with similar content to *Smith* but with more comprehensive scope, dates back to 1964. The list of editors and authors of the latest edition of *Gorlin* is a who's who of modern dysmorphology — names like Raoul Hennekam, the eminent Dutch dysmorphologist; Di Donnai, doyen of British dysmorphologists; the two great Canadian Judiths, Hall and Allanson; and Americans including Bob Gorlin himself, M. Michael Cohen, and others.

As a trainee and in my early days as a geneticist, I would often turn to *Gorlin* in particular. Fresh from seeing a patient in clinic, sure that I had seen his face somewhere before, I'd hunt for syndromes that might match. I'm not a maker of New Year's resolutions, but the closest I've come is the occasional decision that I really should devote some time each day to reading *Gorlin*'s entry on a different syndrome,[2] trying to memorise the details so I'll know the condition when next I see it. If I had ever been able to sustain the effort for more than a few days, I'd doubtless be a better dysmorphologist than I am.

When people started formally describing and naming syndromes, they did not think of themselves as dysmorphologists. None of them were clinical geneticists, because ours is a new specialty, and people have been describing syndromes for a long

---

2   One of these days I'll read Proust's *Remembrance of Things Past*, too. Definitely.

time. Identifying a 'new' syndrome is really an act of discovery, of something that has always been there in the population. As Ecclesiastes says, there is nothing new under the sun. It's true that, before modern medicine, babies born with many of the more severe syndromes would not have survived past infancy, so some of the patterns we are recognising now have only recently emerged; they were always there in potential, but not in reality. But the record of human observation of syndromes goes back many centuries. Two ancient Egyptian gods, Bes and Ptah, had forms of dwarfism — perhaps they had achondroplasia, like the actors David Rappaport and Peter Dinklage. The forms of these gods were undoubtedly based on familiarity with real people affected by (mostly genetic) causes of short stature.

Later art contained ever clearer depictions of specific conditions, as painting techniques improved. There have been attempts to see the features of Down syndrome in the faces of various angels and infant Christs from the 15th and 16th centuries; most of these seem to me to be a bit of a stretch. But in the 17th century, the great Spanish painter Diego Velázquez produced at least ten paintings of people with short stature. Some, such as the *Portrait of Sebastián de Morra* (who undoubtedly had achondroplasia), are clear enough depictions that a diagnosis can be made with confidence, across the centuries.

More recently, the American painter Ivan Le Lorraine Albright's 1929 painting *Among Those Left* unmistakably shows a man who has a condition called Noonan syndrome — 34 years before the cardiologist Jacqueline Noonan's first systematic description of the condition.

People with Noonan syndrome tend to be shorter than average. Many are born with heart problems, particularly a narrowing of the main valve between the right-hand side of the heart and the pulmonary artery (the artery we met a few pages ago, carrying blood to the lungs to be oxygenated). Some struggle at school, but not all — I know of a doctor with Noonan syndrome. There can be other health problems. But it's the look that really defines Noonan syndrome: drooping eyelids; eyes that slant a little downwards from nose to ears; ears that may be set low and tilted backwards; a broad, even 'webbed', neck. Written down like this, it seems like people with the condition would stand out in a crowd, but most are just normal-looking people who happen to resemble one another. As children grow into adulthood, the features tend to become less distinctive, so that, when I make the diagnosis of Noonan syndrome in a child, I'm not always certain whether one of the parents is also affected.

Some of the things that are different about people with Noonan syndrome are so subtle that you'd never see them if you didn't know to look. Their fingerprints, in particular. There are three main types of human fingerprint: whorls, arches, and loops.[3] Magnified, fingerprints look like a series of parallel ridges in the skin of the finger (or of the toe, but for some reason toeprints never seem to attract much interest). These ridges can form various patterns on the pads at the ends of fingers and thumbs. In an arch, some of the lines come in from one side, form a peak that points towards the end of the finger, and then cross to the other side. In a loop, some of the lines come in from the side (usually the side away from the thumb), come in to or across the middle, and then go back the way they came. In a whorl, some ridges loop back to themselves, forming an oval — although there are some variations, like the double whorl, which looks more like an S.

Take a look at your own fingerprints — you probably have loops and whorls, and you may have an arch or two as well. If they aren't easy to see and you don't have a magnifying glass handy, you could photograph them with your phone and zoom in on the picture (good lighting helps with this). While you're at it, have a look at the lines on your palms, and the creases on the

---

3   Not strictly relevant to human genetics, but I can't let this section pass by
     without mentioning that koalas have fingerprints that are indistinguishable
     from human prints. So if the burglars got in through a high window, and
     the only thing that was taken seems to have been the gum leaves from your
     vase of Australian natives ... the police fingerprint database might not be
     much help in tracking down the culprit.

palm side of your fingers. The study of these lines and creases is called dermatoglyphics,[4] and, particularly when I started in genetics, we used to spend quite a lot of time peering at fingers, counting loops, whorls, and arches as part of the effort to reach a diagnosis.

This was because of the observation that there are some differences in patterns in certain syndromes. People with Noonan syndrome, and some other conditions, have a higher proportion of whorls than the population in general. People with Down syndrome commonly have a single transverse palmar crease: instead of two prominent lines running across the palm at an angle, from just below the index finger, they are likely to have a single line running more horizontally across the hand. Don't be worried if you have a single transverse palmar crease — about 1–2 per cent of people have this on one or both hands, just as a normal variation. Likewise, don't read too much into it if you have a lot of whorls, or none. The link with various syndromes is real enough, but there's so much normal variation that, taken in isolation, the information is pretty meaningless. This hasn't stopped people trying to find meaning, mind you — there have been all sorts of attempts to link dermatoglyphics to intelligence, personality traits, risk of cancer, and so on. Even today, there are still companies that will happily take your money,[5] analyse your fingerprints, and issue you with a report, telling you about your strengths and weaknesses and advising on career choices. Since horoscopes are now available for free online, this may not the best way to spend your money.

As well as the extra whorls, people with Noonan syndrome usually have fetal fingertip pads — little bumps in the middle of their fingertips. These, too, can be a normal variation. Like the whorls, though, in people with Noonan syndrome they are a clue to something that happened during early development,

---

4   From Greek words meaning 'skin' (dermato-) and 'carved' (-gluphikos). It's the study of the carving on your skin — rather poetic.

5   Crossing their palms with silver is no longer required. A credit card will do fine.

in the first months after conception. Fingerprints develop from about the tenth week of pregnancy, and by about the 18th week they are complete. Ultrasound scans have shown that a developing fetus with Noonan syndrome typically has an excess of fluid under the skin around this time, because the system for draining it away doesn't seem to work as well as it should. This leaves permanent marks — the fingertips are a little swollen while the fingerprint ridges are forming, which makes whorls more likely, and a little extra skin grows there and stays as fetal pads. More obviously, fluid can collect at the back of the neck, expanding the skin and leaving the 'webbed' look seen in some people with Noonan syndrome.

So, there can be some real clues written on the hands about events during early development. Nowadays, the most useful of these is probably an *absence* of creases. If we see that a child with a neurological condition has absent or faint creases at the joints of the fingers, it tells us that those joints weren't moving much at the time the creases should have formed. This means there was less movement than there should have been, from very early on — making it less likely that, for example, something happened around the time of birth that caused the neurological problems.

Speaking of absences — *almost* everyone has fingerprints, but there are some unfortunate souls who do not. Sometimes, this can go along with other health problems, but there have been a handful of families described with 'isolated adermatoglyphia'. These are people who are otherwise completely healthy, but do not have fingerprints or the usual creases on their palms. You may think this is a harmless state of affairs, and indeed for much of human history it has been. Better than harmless, even, for the occasional criminal with adermatoglyphia who may have been active (and undetected) in the years since the first forensic use of fingerprints in the 1890s. However, it turns out that adermatoglyphia is increasingly a problem when people try to cross borders. More and more countries require a fingerprint on entry. Can you imagine trying to explain to a

US immigration officer why you don't have any fingerprints? Affected families call the condition 'immigration delay disease', for good reason.

Adermatoglyphia is a good example of a descriptive syndrome name. The name describes the condition, in this case concisely and completely: the prefix a- means that something is absent, and in this case it's the dermatoglyphics. It would sound a bit cryptic if you heard it without knowing what dermatoglyphics are, but, now that you do, the word is perfectly straightforward to understand. There are plenty of other syndromes with more or less descriptive names along the same lines. Fibrodysplasia ossificans progressiva means:

fibrodysplasia — an abnormality of fibrous tissue (ligaments and tendons)

ossificans — turning to bone

progressiva — progressively.

If you have fibrodysplasia ossificans progressiva, your fibrous tissue progressively turns to bone — gradually immobilising you. It is an unfortunate and disabling condition.[6]

It's no longer compulsory for medical students to know Latin and Greek, and there are plenty of more straightforwardly descriptive syndrome names. Doctors love an acronym, and geneticists are no exception. VACTERL, for instance, is the association of abnormalities of the Vertebrae (bones that make up the spine), Anal abnormalites, Cardiac (heart) anomalies, Tracheo-oEsophageal fistula (an abnormal connection between windpipe and gullet; the condition was named by an American, for whom the structure is spelled Esophagus), Renal (kidney) abnormalities, and Limb abnormalities.

Sometimes, the urge to make up a clever acronym can lead to unfortunate outcomes. SHORT syndrome includes short stature as one of its features, but it must be a considerable nuisance for the parents of an affected child to explain that

---

6   There are some potential treatments for FOP undergoing trials. Fingers
     crossed that one or more will work.

SHORT syndrome is a real thing and doesn't just mean the kid is small. CHILD syndrome must also be irksome to families. PHACE syndrome does indeed affect the face, and it seems a bit harsh that the syndrome's name draws extra attention to it in this way. SeSAME, CRASH, and RIDDLE syndromes are all a bit problematic, especially the latter (increased Radiosensitivity, mild ImmunoDeficiency, Dysmorphic features, and LEarning difficulties). Nobody wants to hear that their child, despite having a diagnosis, is still a riddle. ANOTHER syndrome[7] would just be a nuisance to explain. CHIME and TARP maybe aren't so bad. But probably it's best, if you're creating an acronym for a new syndrome, to aim for something without an existing meaning: CADASIL, CODAS, GAPO, MASA, MEDNIK, and MELAS all seem pretty safe.

It's not just the names of syndromes that can be problematic. Take the single transverse palmar crease — a single line across the width of the palm, mentioned above. The old name for this, and one that some people still use, is 'simian crease'. Simian, in this context, means monkey or ape-like. Monkeys and the other apes (besides us, that is) don't all have the same pattern on their hands, but some do indeed have a single crease. That doesn't mean it's okay to describe a child by comparing her with a monkey.

Some syndromes wind up with a plethora of names. The champion in this regard is probably velocardiofacial syndrome (VCFS — a harmless enough acronym). People with VCFS may have cleft palate ('velo' refers to the palate), heart abnormalities ('cardio'), and a distinctive facial appearance, which is often very subtle, and certainly not abnormal. Other past and present names for VCFS include:

- 22q11.2 deletion syndrome — for the tiny chunk of chromosome 22 that is missing in most people with VCFS

---

7   I am not making this up.

- 22q11.2 **micro**deletion syndrome — because … it's a *tiny* chunk of chromosome
- monosomy 22q11.2 — another way of saying that there's a tiny chunk of chromosome 22 missing
- CATCH22 … never mind why, but suffice it to say that this is another prime example of how *not* to construct a medical acronym. This was graciously acknowledged by the originator of the term, Professor Sir John Burn, the great British geneticist, when he formally withdrew the name from use in 1999, just six years after its introduction. As he noted, in Joseph Heller's book, the titular Catch-22 describes a no-win situation[8] — again, not something you want applied to yourself or your child.
- CATCH phenotype — a short-lived attempt to retrieve the CATCH part of the CATCH22 acronym, in the face of criticism by patient groups.
- conotruncal anomaly face syndrome (CAFS) — 'conotruncal' refers to a particular group of heart malformations seen in babies with the condition; 'face' because of the characteristic facial appearance.
- DiGeorge syndrome
- Sedláčková syndrome
- Shprintzen syndrome
- Strong syndrome
- Takao syndrome.

There's a good reason why the condition has so many different names. It's relatively common, but it can show itself in

---

8 You've probably read *Catch-22*. But just in case: the catch was that, for the World War II pilots in the book, if you're crazy, you don't have to fly missions. But asking to be grounded was proof of rational thinking … so you couldn't be crazy. In the words of the book, 'Orr would be crazy to fly more missions and sane if he didn't, but if he were sane he had to fly them. If he flew them he was crazy and didn't have to, but if he didn't want to he was sane and had to.'

different ways, and so was described independently by various different people. Later, we realised that each of these apparently different conditions was just another version of the same thing. Even now, there's an argument that it may be worth keeping some of the different names. Angelo DiGeorge, an American paediatric endocrinologist, described children affected by a particularly severe form of the condition, with a life-threatening immune deficiency. Most people with VCFS don't have problems with their immune systems. You might be able to measure some differences in their white blood cells, but they don't have more infections than anyone else. So, maybe if we see a child with a deletion of the right part of chromosome 22 who also has a severe immunodeficiency, we should still speak of DiGeorge syndrome.

Otherwise, though, it's generally best if a genetic syndrome has a single name. This has a number of advantages — it makes it easier for patient groups to find one another, it unifies the medical literature, and it prevents confusion.

Of the various names of people listed above, DiGeorge and Shprintzen are the two best known among geneticists. For a while in the 1990s, Shprintzen syndrome looked like it might wind up settling down and becoming *the* name of the condition. In 1978, Dr Robert Shprintzen — a speech pathologist — and colleagues published a description of 12 children with the condition that they named VCFS. Three years later, the group published another paper describing 39 individuals and correctly identifying the inheritance pattern of the condition. Over the following years, Shprintzen made contribution after contribution to our understanding of VCFS, to the point that others in the field started referring to it as Shprintzen syndrome. But as far as I can tell, Shprintzen himself never used the name, always using his original term — velocardiofacial syndrome.

Had the name stuck, it may have been something of an injustice, through no fault of Robert Shprintzen. In 1955, Eva Sedláčková had pre-empted Shprintzen by more than two decades, publishing a description of 26 patients with — as it later emerged — the same condition. Her paper was written in Czech, so it was hardly surprising that Shprintzen and colleagues had no knowledge of it. At the time, Prague was an important European centre for the treatment of cleft palate, and the term Sedláčková syndrome found its way into textbooks in German and French over the coming years, but it wasn't until the 1990s that the link with VCFS was firmly established, and Sedláčková's achievement came to the notice of the English-speaking world (or at least, that part of the English-speaking world with an interest in velocardiofacial syndrome).

Getting a syndrome named after yourself can be a bit of a lottery. Quite often, things work out fine — someone is the first to describe a particular syndrome, the next person to publish a paper about the condition acknowledges this by attaching the original describer's name to the condition, and all is well. However, there have been plenty of instances where things are not so straightforward — possibly we should all be doing our best to pronounce Dr Sedláčková's name every time we refer to VCFS, for instance. It's also not always obvious in retrospect why a particular name or combination of names is used — sometimes just the first author of an original paper, sometimes more than one author, and so on. There are some charming eccentricities. Why, for example, does Cornelia de Lange syndrome — the only instance of this that I know — include the first name of Dr de Lange, instead of just her surname as is usual?

Sometimes, two or more doctors have their names joined together in the naming of a syndrome. Occasionally, this happens because someone has been particularly prolific, and the use of an additional name avoids confusion. There's a Shprintzen-Goldberg syndrome, which could not have been named just Shprintzen syndrome because the name was already taken. Confusingly, there's also a Goldberg-Shprintzen syndrome; same Goldberg, same Shprintzen, completely different conditions. Perhaps the champion of multiple genetic syndrome names is John Opitz. His name is attached to Opitz G/BBB syndrome (also known as Opitz G, Opitz BBB, and Opitz-Frias),

Opitz C syndrome (also known as Opitz trigonocephaly syndrome), Opitz-Kaveggia syndrome, Bohring-Opitz syndrome, and Smith-Lemli-Opitz syndrome. The G, BBB, and C come from the practice Opitz had in the 1960s and 70s of naming newly described syndromes after the first initials of the patients he was describing (a typical paper was entitled 'The C Syndrome of Multiple Congenital Anomalies'). This never really caught on among others in the field, and wouldn't have been useful much past 26 syndromes in any case.

Shprintzen's modesty regarding the term Shprintzen syndrome is not unusual, by the way. There were (in retrospect) some reports of Noonan syndrome well before Jacqueline Noonan's first full description of the condition, and it seems it took a long time for her to become comfortable with her name being used in this way — it was nearly ten years before she would write a paper in which she referred to Noonan syndrome. In a special issue of a journal published in 1968, four of five papers about the condition had a running title 'Noonan syndrome'; the odd one out was written by none other than Dr Noonan.

This modesty is not just something from a bygone age. In 1998, my friends David Mowat and Meredith Wilson, with colleagues, described a syndrome that others soon dubbed Mowat-Wilson syndrome.[9] Although David and Meredith eventually published papers that use the name, I don't think I've ever heard either of them utter the words 'Mowat-Wilson syndrome'. The closest they come is to use the acronym 'MWS'.

Of course, wherever there's a trend, there's someone who

---

9   I once made the diagnosis of Mowat-Wilson syndrome in a child I saw in a shopping mall. I was going down a moving ramp; she was coming up the other side, in a wheelchair. She had absolutely typical facial features. There was no doubt in my mind about the diagnosis. For a few seconds, as we approached each other, I wondered — should I speak to the family? It would feel rude and intrusive, but what if they didn't know their child's diagnosis? I couldn't decide, and my indecision became a decision. The family went past me, and out into the car park, and the opportunity was lost. I will never know if I did the wrong thing.

will buck it. You have to admire the chutzpah of Drs Bettex and Graf, who decided to cut out the middle-man in entitling their 1998 paper 'Oro-palatal Dysplasia Bettex-Graf — a New Syndrome'. Sadly for them, the term seems not to have caught on, or perhaps the condition is just too rare for others to have seen it and considered adopting the name.

There have been occasions when the naming of syndromes has generated controversy. Sometimes, this relates to disagreements about precedence, or just national pride. Is it Silver-Russell syndrome or Russell-Silver? Silver was American, Russell British ... guess where the term Silver-Russell is more common? More seriously, we no longer refer to Hallervorden-Spatz disease — now we call it pantothenate kinase–associated neurodegeneration (PKAN for short).

This is not because of any dispute about priority. In a 1922 paper, Julius Hallervorden and Hugo Spatz described five sisters affected by a condition that had never been reported before; the paper contained considerable clinical and pathological detail. Hallervorden was a pioneering neuroscientist, described by a former student as having 'a quiet, reserved nature' and being 'wholly devoted to science and to neuropathology, and at the same time, warm, friendly and an inspiring teacher'. Others described him as 'good-natured, personally modest and possessing a dry humour'. During his career, Hallervorden published 120 papers and received many honours for his contributions to medical science. So — Hallervorden was an all-round nice guy, an eminent scientist, and had clear priority in describing what proved to be an important neurological condition. The perfect person to honour with an eponym, surely?

Except ... except for the source of some of his scientific material. The work that led to the first description of PKAN was conducted in Munich, at an institute led by Hugo Spatz. In the late 1930s and 1940s, Hallervorden was working in Berlin. At the time, the Nazi doctrine of 'racial hygiene' was being enacted, in the form of the systematic murder of children who were regarded as mentally defective, or were physically disabled by malformations; and in the form of the T4 program, in which more than 70,000 people taken from psychiatric hospitals were gassed to death. Hallervorden evidently saw all this as a wonderful opportunity to obtain material for study. In 1942, he wrote, 'I have been able to dissect 500 brains from feeble-minded individuals'. Hallervorden was fully aware of the source of the brains he studied. He is quoted as saying, 'I heard that they were going to do that, and so I went up to them and told them, "Look here now boys, if you are going to kill all those people, at least take the brains out so that the material could be utilized"'. When asked how many he could examine, he replied, 'I told them an unlimited number — the more the better'. There's more to the story than this, and worse — but you get the picture. PKAN it is.

Among eponymous genetic syndromes, the most famous is Down syndrome.[10] John Langdon Down was a British doctor who became chief physician of the Earlswood Asylum in the early 1860s. Down seems to have done a fine job, transforming a brutal and filthy institution into one in which the inmates were treated with kindness and humanity. Down took the opportunity to study his charges, and in 1866 published his famous paper 'Observations on an Ethnic Classification of Idiots'. Just a few pages long, it was to stamp Down's name in medical history. While we would not be comfortable with much of Down's language today, the paper makes fascinating reading and is well worth seeking out. Down starts by pointing out the

---

10   Not *Down's* syndrome, by the way. Although the possessive in syndrome
     names lingers on, it has been considered incorrect since at least the early
     1980s. Down didn't own the syndrome, nor did he have the condition
     himself. Perhaps the only eponym that still deserves its possessive is
     Trousseau's sign. Trousseau, a 19th-century French physician, noticed
     that people with cancer sometimes have episodes in which clots form
     in blood vessels under the skin and cause inflammation; the problem
     moves from place to place on the body, and is thus known as migratory
     thrombophlebitis. The unfortunate Trousseau diagnosed his own stomach
     cancer when he developed migratory thrombophlebitis, making this truly
     'Trousseau's sign'.

difficulty of classifying 'congenital mental lesions', as he calls them — a problem we are still wrestling with more than 150 years later. He criticises existing systems of classification and then canvasses some possible causes one might need to consider, especially environmental influences, addressing the question of whether the problem arose before or after birth. 'Has the nurse dosed the child with opium? Has the little one met with any accident?', and so on.

Down points out that 'the maternal imagination' might have identified a number of possible causes, which may or may not be relevant. This rings true today: it's very common that the parents of a child born with a malformation, or diagnosed with a disability, wonder whether some incident during the pregnancy was responsible — be it stress, occupational exposures to chemicals, or fumes from painting the nursery. One of the most useful things I can do is to provide reassurance about this, because it's very seldom that the source of concern is actually the cause of the child's problems.

Where Down goes off the rails, from a modern perspective, is in the way he classifies the different conditions he sees: by race. He identifies 'examples of the Ethiopian variety' that he calls 'white negroes, although of European descent', and others of the 'Malay variety'; at this remove, it's not clear which conditions he is referring to. However, it is very clear what he means when he refers to 'the great Mongolian family', because the term 'mongolism' for Down syndrome was common at least until the 1960s, and probably later. Down saw similarities in the facial appearance of some of his charges with the faces of people of central Asian descent, and concluded that *his patients were in fact Asian*. Charmingly, he goes on to conclude that, if this is the case, the differences between the races are only superficial, and that his observations 'furnish some arguments in favour of the unity of the human species'.

With 150 years of scientific progress between us and John Langdon Down, it is easy to be critical of his premises, and his conclusions. But his famous paper (and subsequent writing) includes some useful and accurate clinical information. As he correctly observes, Down syndrome is present from birth, and never results from accidents after birth. He describes some physical characteristics, and also features of the personality that often accompanies the condition: 'they are humorous, and a lively sense of the ridiculous often colours their mimicry'. He points out that affected individuals are usually able to speak, and he makes the point that, given the opportunity, they can learn manual skills. Down's is a fundamentally humane approach to his subjects, a standard that has not always been met by his successors.

In non-human biology, there is a rigid rule about naming new species: priority is everything. The first to describe a species gets naming rights. This rule dates back to the efforts of Hugh Edwin Strickland, who in 1837 proposed a set of rules in an attempt to reign in a then-chaotic situation. At the time, the naming of species was in a parlous state, with people attempting to rename species and many species having numerous different names. Strickland, a naturalist, was spurred into action when he read a 1934 proposal to rename the bullfinch to 'coalhood'. Black though the head of a bullfinch may be, this was the final straw for Strickland, who at once began work on an initial set of 22 rules, with primacy of description taking a prominent place.

It took several years for Strickland to persuade the British Association for the Advancement of Science to adopt the rules, and longer for the scientific community to adopt the standard. Later, Strickland's rules evolved to become the current International Code of Zoological Nomenclature;[11] for well over a century, there have been clear rules in place, not without occasional controversy regarding interpretation but nonetheless accepted by all[12] — even when the rules lead to decisions that those of us who are not zoologists might find unfortunate. You may

---

11   There is also an International Code of Botanical Nomenclature.

12   Sadly, Strickland did not live to see his triumph. He died in 1853, aged only 42, in a freak railway accident. He had gone to inspect geological strata exposed by the construction of a railway cutting, through which parallel train tracks ran. Seeing an oncoming freight train, he stepped aside onto the other line — only to be hit by an express travelling in the opposite direction.

have heard, for instance, that the wonderfully-named dinosaur *Brontosaurus* ('thunder lizard') was deemed to be only a variant of *Apatosaurus* ('deceptive lizard', although it's hard to see how deceptive an animal could be when it had an average length of over 20 metres and weight of around 20 tonnes). The name *Brontosaurus* was scrapped because *Apatosaurus* was described first, a blow to amateur dinosaur-lovers everywhere.[13] Those of us who grew up dreaming of being palaeontologists rather than fire fighters may have found this a bitter pill to swallow, but at least there is a consistent set of rules.

The current system for naming syndromes has a certain charm, despite its capriciousness. But perhaps we in genetics should join the rest of biology in taking a more systematic approach to choosing names.

Regardless of how they come by their names, there are an awful lot of syndromes. This means that beyond a core set of conditions, plus some extras depending on the skill and experience of the dysmorphologist, if you want to do dysmorphology with success, you are going to need help. That help can come in the form of textbooks, like *Smith* and *Gorlin*. It can come in the form of searchable databases, like the London Dysmorphology Database and the Australian database, POSSUM — Pictures of Standard Syndromes and Undiagnosed Malformations. These can be very useful. As a new trainee in clinical genetics, the very first time I was asked to see a child with an unexplained syndrome, I came back from the intensive care unit to the genetics department, plugged the child's clinical findings into POSSUM, and the correct diagnosis came up. 'Dysmorphology is easy!' I thought. It was *years* before that happened again.

More recently, increasing efforts have been made to use modified facial-recognition software to match patient photographs, including 3D photographs, to databases of patient photographs. The technology is starting to get quite good — just

---

13 A 2015 paper may have resurrected the name, arguing that *Brontosaurus* truly was a different beast from *Apatasaurus*. It appears the matter will not be settled until more fossils are discovered.

as other technology is coming along that will probably make it redundant (as we saw in chapter 5).

But in the end, the dysmorphologist's best source of help is often other dysmorphologists. On Monday afternoons, the members of our department get together and discuss every patient we've seen in the past week. This review meeting is essential not just for dysmorphology but for all the rare situations we deal in; every patient gets a virtual second opinion (and third, fourth, and fifth). We show photographs of the patients we see: faces, hands, feet, birthmarks, whatever we think might be a clue. Dysmorphology has its own jargon — in describing the shape of the ear alone, we might speak of differences in the helix, the antihelix, the crus, the tragus, the concha ... the list goes on. In 2009, Alasdair Hunter and colleagues published a 21-page paper, with over 70 figures, explaining how a dysmorphologist should describe the ear.

This was just one in a series of papers, the 'elements of morphology'. We talk about handles — yes, I know you can grip someone by the ears, but these are metaphorical handles; features distinctive enough that they are likely to be helpful in reaching a diagnosis. An ear that's a bit small isn't much of a handle, because it's such a common thing, including in people who are otherwise completely well. An ear that looks crumpled, as though it's been squeezed in someone's fist while developing — well, that's a decent handle, and might be a big part of what leads you to a diagnosis.

As is true for any skill, some people are better dysmorphologists than others. Fortunately for me and for my patients, I work with several people who are very good at it indeed. There have been a number of occasions over the years when I've come from clinic without a clue about a particular patient, and someone at that afternoon's review meeting has made a suggestion that led to the correct diagnosis.[14] In particular, my colleague

---

14   Only once or twice has it gone the other way — but I remember and treasure those exceptions.

Rani Sachdev is a diagnostic bloodhound; give her a sniff of a diagnosis, and she won't give up until she runs it to ground. Many a time I've come to work on a Tuesday morning to find an email from Rani, sent in the middle of the night, with a couple of journal articles attached and a message saying, 'Ed, I was thinking about your patient from review yesterday and I was wondering about (syndrome X). Have a look at these papers and see what you think.'

Even with all that expertise in the room, and even with textbooks, databases, and online resources ... until recently our hit rate was pretty terrible. So when you think there's a diagnosis to be made, but you can't figure out what, and your colleagues can't figure it out either (even Rani) — then perhaps it's one for the Dysmorphology Club.

This is undoubtedly one of the world's oddest clubs. You can't apply for membership, but only members may attend. It follows that you also can't resign, although, since attendance is voluntary, and a privilege, you can always stop coming, if you choose. The club meets twice a year. Its rules are few. Be brief. If you bring an unknown (pictures of a patient in whom you are hoping for a diagnosis), you should bring a known (pictures of a patient in whom you have made a diagnosis, preferably of a rare and little-known syndrome). Most importantly, respect the privacy of those who are not present, but who are essential to the club's success: the patients and their families.

The proceedings of the club are always the same. We meet once a year at the annual scientific meeting of the Human Genetics Society of Australasia, and once in a standalone meeting somewhere in Australia. The attendees are clinical geneticists and trainees in clinical genetics; otherwise, the meeting is closed. There is a fairly strict limit of two patients per person, although, since trainees speak as well, you may be able to sneak extras in via your trainee's allocation. Time is tight, but not everyone is good at being concise; the key features of the patient's family history, medical history, examination findings, and previous investigations are summarised, and photographs are shown, all with the consent of the families or individuals concerned. The aim is to educate your colleagues and to get clues to a diagnosis for your patients. Instead of the handful of extra opinions at review meeting, you're getting scores of expert eyes and minds on the case, often including a visiting international expert. It's a great institution.

These, then, were the tools available to me in trying to make a diagnosis in Dianne. In the end, however, I made the diagnosis using a different set of tools altogether: laziness, and luck.

Before digital photography, we would take two cameras to clinic with us — a Polaroid Instamatic and a standard camera. The former was so that we would have at least some photographs available to pass around at review meeting, the latter so that we had higher quality pictures for the patient's file. At least once a week, we'd go up the road and drop rolls of film off to be developed. The prints would be distributed to the relevant geneticist or trainee, who would then make sure they were put in the correct file.

Or, if the geneticist was indolent or a procrastinator, he would just leave the photographs lying on his desk, awaiting the day that he got round to doing something about them. So it happened that Dianne's photograph was sitting smiling up at me on top of a pile of paperwork, for ... quite a few months. When you look at someone's photograph several times a day for months on end, you wind up knowing exactly what they look like, which in this case proved to be a good thing.

One day I went to the room where we kept the various genetics journals we subscribed to, in order to look up an article on a completely different topic. I was leafing through the relevant copy of the *American Journal of Medical Genetics* — and suddenly, there was Dianne, staring back at me. The child in the article looked like her twin! Excitedly, I read the article, which described the condition that came to be known as Cantú syndrome. Affected children were born prematurely, weighing more than expected for their gestation. They had an excess of hair, large hearts, often with a patent (open) ductus arteriosus, and a distinctive facial appearance. It was all an exact match with Dianne's problems. There was no doubt in my mind that this was the answer.

At the next Dysmorphology Club, I showed pictures of Dianne as a known diagnosis, expecting that I would stump my assembled colleagues.[15] There had only been a handful of people with Cantú syndrome described in the medical literature; the name wasn't even firmly established yet. To my surprise, Stephen Robertson — a New Zealander — spoke up from the back of the room, correctly identifying the diagnosis. He had recently made the same diagnosis in one of his patients, using the more traditional approach of applying actual diagnostic skill, rather than relying on the approach that I had used. Steve and I went on to write a paper together, describing our patients.

What did this diagnosis mean for Dianne, and for her family? At first, not as much as you might think. There were very few children who had been described with the syndrome, and we were learning far more from Dianne and from the others we started seeing at the time than the families could learn from us. As a result, Dianne has featured in four different scientific papers. Still, there was some useful information we could give her parents. Dianne had been developing a little slowly, and in particular she was slow to sit and to walk. We knew already that this was a common pattern in children with Cantú syndrome, but that at least some of them went on to have normal development and normal intelligence.[16] This meant we could provide some cautious reassurance about Dianne's future.

Parents of a child newly diagnosed with a syndrome typically are concerned to know what caused the problem, and

---

15 Okay … dysmorphology club does have a third goal. Showing off.

16 Later, we were to learn that *most* people with Cantú syndrome have normal intelligence. I've met a psychologist and a doctor with Cantú, and we now know that intellectual disability is the exception, not the rule.

whether it might happen again if they have more children. Until then, it had been thought that this was an autosomal recessive condition: that both parents needed to be carriers, and that the future siblings of an affected child would have a 1 in 4 chance of also being affected. When Steve and I were writing our paper, we read the whole of the medical literature about the syndrome — not that difficult, as ours was only the sixth paper published. Only 12 people had previously been described. The reason this had been thought to be a recessive condition was that in the first paper, describing the syndrome, two affected children had been born to unaffected parents. In another family, the parents of an affected child were cousins — a pointer to recessive inheritance, because we share genes with our relatives (and thus are more likely to share the same faulty gene).

By 1997, Cantú and his group were expressing uncertainty about this, because no more families had been seen with more than one affected child. When we wrote our paper in 1998, I was able to count 39 brothers and sisters of people with Cantú, only one of whom — from that very first paper — had the condition as well. We would have expected about nine or ten, a big difference from just one, and strong evidence that recessive inheritance was unlikely. We now know that this is an autosomal dominant condition — only one of the two copies of the gene needs to be faulty to cause Cantú syndrome. Sometimes, the faulty gene is inherited from an affected parent, who in turn may have inherited it from one of their parents; sometimes, the change has arisen anew in the egg or sperm that joined to form the affected child. So how could two children be born to unaffected parents? Almost certainly, this was due to one of the parents of these children being a gonadal mosaic (as you'll recall, mosaicism was described in chapter 5).

The story of Cantú syndrome over the 20 years that have followed Dianne's first appearance in the scientific literature has been one of international collaboration and friendship, of dazzling scientific insight and slow, patient progress. In 2006, an American geneticist, Kathy Grange,[17] made the inspired observation that people with Cantú syndrome have a lot in common with people who have been treated with a drug called minoxidil. Minoxidil is now rarely used for its original purpose, treating high blood pressure, but is still very popular (in cream form) because it makes hair grow, a salve for the balding. Kathy somehow made a link between the side effects of an increasingly obscure treatment and a genetic condition.

If you think of television's version of the brilliant diagnostician on television, who comes to mind? House, perhaps. The irascible Perry Cox from *Scrubs*, maybe, or George Clooney's Dr Ross in *ER*. By and large, the character is assertive, often eccentric, and (apart perhaps from some of the characters in *Grey's Anatomy*) has almost always been male. Kathy — a softly spoken, calm, kindly woman — could hardly be farther from this stereotype. But brilliant she undoubtedly is.

In 2012, two Dutch groups, both working with doctors from around the world who had seen people with Cantú syndrome, independently discovered a link between a gene called *ABCC9* and Cantú syndrome.[18] Dianne was one of the patients who helped with that discovery (her third appearance in a medical journal). What, exactly, had they found?

The body's cells are surrounded by a fatty membrane. When you see drawings of a cell, this membrane typically looks like a smooth, unbroken surface. Really, it is absolutely *carpeted* with proteins, which do many different jobs. Some of them are like radio receivers, listening out for messages from elsewhere in the body. Some of them join one cell to the next (without them,

---

17   Known in her scientific publications and email address as Dorothy K. Grange, but Kathy to her friends.

18   This kind of thing happens remarkably often in science. It's not at all unusual for a journal to publish two papers in the same issue, with different groups reporting the same discovery. Sometimes, the simultaneous publication is no coincidence — each has become aware of the others' work, and they have coordinated their submissions in order not to scoop each other. Nonetheless, for there to be any such agreement, the starting point had to be that both had made the discovery at the same time.

you'd disintegrate in a gooey mess). And others regulate the levels of various substances on either side of the membrane — also vital to your survival. Much of what happens inside cells needs the levels of salts and acids to be tightly controlled in order to work. Your nerves and your muscles, including your heart, can't function without the flow of calcium, sodium, and potassium across the cell's membranes being just right. Stop this flow from happening altogether and you'd be dead in seconds. Mess up just one of the many, many different channels, and the effects vary from nothing at all, through to problems like sudden death or severe childhood-onset epilepsy.

One particular type of that last group of proteins is a channel in the cell membrane that helps to control the flow of potassium ions across the membrane. That channel can be open — letting the potassium flow — or closed, depending on the needs of the cell. In people with Cantú syndrome, the channel is jammed into the open position, allowing a constant, uncontrolled flow of potassium across the membrane.[19] We don't fully understand how this links to all the various features of Cantú syndrome — for example, we have only guesses about why this should cause hair to grow excessively.

It turns out that minoxidil works by jamming this exact channel open, closely mimicking the situation in Cantú syndrome. Kathy Grange was spectacularly right.

A diagnosis is a powerful thing. It gives answers, and it helps with the future — with prognosis and with treatment, and in allowing couples to make informed choices about future pregnancies. Although the dysmorphologist's role has been changed by the new sequencing technology, that human element remains important — for choosing the best tests, and especially when there is uncertainty about the meaning of the results. No matter how cheap, fast, and easy to access genomic testing becomes,

---

19   In turn, this likely suppresses the electrical signals that would normally be happening in affected tissues. Genetics is simple, but cell biology and physiology — the way the body's systems work together — can be complex!

we will still need people with the skills of a Rani Sachdev or a Kathy Grange, and there will still be times when our best chance of getting answers is by taking the question to the Dysmorphology Club.

8

How to make a baby

All you need is love

JOHN LENNON

There's a slogan popular among those opposed to same-sex marriage: 'God made Adam and Eve, not Adam and Steve'. Increasingly, around the world, Adam and Steve *can* get married. They can even have children — but only with the help of someone to provide an egg, and someone to carry the baby. We're nowhere close to developing an artificial womb, but could Adam and Steve dispense with the egg donor, and have children to whom both are biological parents? Equally, could Anna and Eve skip the sperm donor?

The answer to that question may soon be yes.

The question might also be posed this way: could Adam make an egg? Could Eve make a sperm?

Which leads to the question: what *are* eggs and sperm, anyway? And could we make them without the aid of ovaries and testicles? Surprisingly, the answer lies in part with a little boy, hovering on the edge of death.

James was just three when, one day, all of his muscle cells fell apart.

He had been a happy, healthy boy, always on the go, an explorer. Like any small child, he had had his share of colds and other minor illnesses, but nothing out of the ordinary. Then he ran into a virus that stressed his body just slightly more than the others had — and uncovered a vulnerability that had always been there, but had never before revealed itself. James woke one morning cranky and irritable, complaining of pains in his legs. Then he collapsed. His mother called an ambulance, which only just got him to the hospital in time.

Your muscles contain an enzyme called creatine kinase, shortened to CK. It's an important enzyme, but for this story it doesn't really matter what it does — just that muscle cells contain a lot of it, and, when they die, CK is released into the bloodstream. Muscle cells wear out and are replaced all the time, so there's always a little bit of CK in your blood. Normally, that would amount to less than 200 units of enzyme per litre — more if you've just run a marathon, but not usually a lot more. The first time I heard about James was from the intensive care specialist who was looking after him. He was marvelling at James's CK. '500,000! It must be a world record!' And he thought the problem must surely be genetic, which was why I was there.

Record or not, it was bad news for James.

The problem did turn out to be genetic — James had two faulty copies of a gene called *LPIN1*, which is important for keeping the lining of muscle cells stable and strong. In its absence, his muscle cells were fragile, needing only a small push — for example, from what would otherwise be a mild viral infection — to make them break down completely. The same virus that might have given you a runny nose and some aches and pains was potentially a deadly threat to James.

When cells die, other substances pour into the blood. Sometimes, the effects of this can be fatal; cells are rich in potassium, but too much of it in the bloodstream can stop your heart. James had escaped that fate, but he was still in terrible trouble. His muscles felt hard to the touch, like wood wrapped in rubber. We worried that he might have a problem called compartment syndrome, in which swollen muscles are trapped inside their sheaths and the resulting rise in pressure blocks off their own blood supply, so that the muscle dies completely. The surgeons decided to do an operation to check for this, and if necessary release the pressure in his leg muscles. After a single cut, they closed him up again, sure there was nothing to be done. The muscle looked ghastly — pale, bloodless, and seemingly beyond hope.

Without muscles, you can't walk or use your arms and hands, and you can't breathe. James was totally paralysed, and most of us thought he was going to stay that way. Only one of the team looking after James was optimistic about his chances. His neurologist, a man of long experience in the field, told us, and his parents, that he thought there was every reason to be optimistic, and that James could make a full recovery. I found it hard to believe, and felt cross with my senior colleague for giving the parents false reassurance.

Six months later, James walked into my consulting room looking as though he had never been sick.

How was this possible? Although many cells in each of his muscles had perished, some had survived. And among the survivors were some that were special.

They were stem cells.

Sperm are relatively simple critters. They are little guided missiles, carrying only the bare essentials needed to get to their targets. In the head of a sperm is the payload — half of the man's genome (he's only potentially 'the father' at this point). There's a mid-section full of mitochondria — that's the power plant. And there's the tail, a thrashing motor driven by that power plant. A sperm has one job, and one job only: get the payload to its target. When a man ejaculates, there are generally hundreds of millions of sperm released. Most of the time, there is no egg waiting for them, and they die, unmourned and unfulfilled. But even if there is an egg, only *one* of all those millions can fertilise it. A man might produce 500 *billion* sperm in his lifetime, with only one or two fulfilling its destiny. If you were naturally conceived, the sperm that makes up half your DNA was the winner of a contest against incredible odds. The same goes for each of your parents, and their parents, and so on. If you ever find yourself lying awake at night wondering if there is anything special about you, consider this — you are the culmination of a long, long line of outstandingly lucky sperm, stretching back over many millions of years. They were all great swimmers, to be sure — but mostly they were freakishly fortunate. There are many, many ways for a sperm to fail to reach or fertilise an egg. The odds against your existence are genuinely astronomical. You are, without a shadow of a doubt, something special.

Once that magical, one-in-hundreds-of-billions of sperm hits its mark, its job is done. It is swallowed up by the egg, which gathers up the DNA in the head of the sperm and *hunts down and kills* the DNA in the sperm's mitochondria — and then the real work can begin.

To understand how remarkable an egg is, you first need to know the difference between a muscle cell and a liver cell. Why is one of them good at contracting powerfully, while the other would never dream of doing such a thing, but excels at cleaning toxins from your bloodstream, and manufacturing the proteins your blood needs so it can clot when you're bleeding? Both cells have exactly the same DNA — but they use it in different ways.

You can think of the genome as a box full of electrical components, all wired to a single giant circuit board. There's everything you need to make a television, everything you need to make a hairdryer, everything you need to make a microwave oven, and so on. It's said that there are over 200 types of cell in the human body. There's reason to think there might be considerably more than that — there may be important differences between a skin cell on your elbow and one on the tip of your nose, for example. But let's say there are only 200. This means that the box of components in the cell nucleus has everything needed to make any one of 200 different electrical items. There are some things that almost every type of electrical equipment needs — a way of getting electricity from the outlet, for instance — and there's only one of those in the box. Similarly, there are some components that every cell needs, such as the toolkit that is used for disposing of damaged proteins. On the other hand, only your fridge needs a compressor, just as only a particular set of cells in your pancreas needs to make insulin.

Let's say our electrical box is set up so there are switches that select whether any one component is used or not, and let's pretend it doesn't matter in what order they are joined up. This means that, by choosing which of the components in the box to switch on, you can make it behave very differently. The same box could function as a computer, or a printer, or an electric mixer, or a bandsaw. That's pretty much what happens in a cell, too. Each cell has exactly the same set of 23,000 genes, but in each type of cell only a particular set of those genes is switched on — the rest are silenced, more or less permanently. There's a set of genes that are switched on in every cell, known as housekeeping genes; there are genes that are needed by multiple cell types, but not all; and there are some genes, like the insulin gene, that are specific to just one type of cell.

The special thing about the egg is its potential. A fertilised egg can — and *must* — give rise to every other type of cell, including the eventual eggs or sperm that will become the next generation. I like to imagine this first cell as vibrating with energy, bursting with potential. That first cell, and all of its daughter cells for the first few cell divisions, are the ultimate stem cells. They are totipotent — literally, 'wholly powerful' — meaning that each of those cells is uncommitted, and can become any of the hundreds, or perhaps thousands, of cell types needed to make a person — and also the placenta, needed to support and nourish the baby until it is ready to be born. One step down are the pluripotent stem cells, which are 'severally powerful' — although the only thing they can't make is a placenta. Identical twins are evidence of the power of pluripotent stem cells — if something causes an early embryo to split, you get two babies for the price of one.

As the embryo develops, however, cells start to set off down particular paths, gradually becoming more and more committed until most of them have reached their final form (a bit like an evolving Pokémon). Once a liver cell is a liver cell, that's all it will ever be. The molecular switches that choose the particular set of genes needed by that cell type are welded into place. The television will only ever be a television; the blender will never download an ebook.

Except ... except for the cells that *don't* go all the way down that path. In every part of your body, there are cells that didn't quite commit all the way. Most of them sit around waiting to repair damage — these are the ones that saved James, by regrowing his muscles.[1] Some are very active, like the ones that live in your bone marrow making new blood cells, or the ones that replenish the cells in your gut, which are always being worn away by the contents of your bowel. Others seem to sit around for a very long time, ready for when they are needed.

Stem cells can be more or less specialised, too. A haemocytoblast can become any type of blood cell. Once it commits to becoming a megakaryoblast, it's still a stem cell, but all it can ever make is megakaryocytes. These, in turn, are the weird cells that make platelets, the little cellular scraps that circulate in your blood, waiting for the chance to help make a clot. Megakaryocytes — unlike the tiny platelets that they make — are huge, and have an enormous nucleus, with extra sets of chromosomes. While they are forming, they double and redouble their chromosomes, and they can have as many as 32 times as many chromosomes as a cell usually does. We have no idea *why* they do this, by the way.

So — stem cells are just cells that have the flexibility to become other types of cell. Unfortunately, the power of stem cells is limited. Often, damage to tissue leads to a scar, rather than replenishment by stem cells. Still, there is a lot of interest in using stem cells as medical treatments, getting them to repair damaged

---

1   The stem cells in muscle are called satellite cells. When muscle is damaged, they divide; some of the daughter cells remain as satellite cells, ready if they are needed in the future. The rest merge with the damaged mature muscle cells and repair them.

tissue — such as after a heart attack — with healthy new cells. For our purposes, though, what matters is that stem cells have this flexibility, and that the fertilised egg is the ultimate stem cell, ready and able to become every other type of cell.

For Adam and Steve's purposes, we don't need Adam to make a *fertilised* egg — we can get sperm from Steve, after all. But we do need to be able to persuade a cell that has gone down one path in life to change its mind and mature first into the precursors of an egg and then into an actual egg. Making an egg is a complex process, but, if we could persuade any cell type to become another cell type, there's no reason the second cell type couldn't be an egg.

Can we do that persuading? In principle, there's no reason why not.

Professor Richard Harvey, who was my PhD supervisor, is an eminent biologist who studies the way the heart develops. Richard's lab uses a variety of different methods to try to understand the complex processes that lead to the formation of the heart and all its structures. Once, I was visiting the lab and Richard beckoned me over to a microscope and asked if I wanted to see something cool. There's only one answer to that question ... and he delivered, in spades. Looking through the microscope, I saw a glass slide with clumps of cells growing on it. The cells were *pulsating rhythmically.* These were cardiac organoids: clumps of cells that had been persuaded that they were actually a heart, and were behaving as though this were true, pumping away faithfully. They had been made by treating stem cells with cellular messages that said: you will be part of a heart.

Even cooler, Richard's lab had started out by *making the stem cells* from mature skin cells. In other words, it is already possible to make stem cells by starting with mature, fully com-mitted adult cells and hitting reset. What you get are known as induced pluripotent stem cells — iPSCs. That sounds like a mouthful, but all it means is cells that have been chemically persuaded to roll back the clock, to the time when they were cells that could become any other type of cell.

Could you use the same method to transform a skin cell into an egg? There's no reason why not, in theory at least. In fact, quite a lot of progress has been made towards producing sperm (which are easier) from stem cells. The reason given for attempting this has been as a possible treatment for infertility, but, once it's possible to use skin cells from a man to make sperm, it might not be too hard to use cells from a woman to achieve the same thing. Eventually, the skin-to-egg method is likely to be possible as well, although making an egg is a much harder task. Anna and Eve might get first crack at this technology, it seems. Still, the technical challenges are just that — challenges, which can undoubtedly be overcome.

However, it's likely that there will be objections raised to actually making sperm from a woman's cells or eggs from a man's. From a practical perspective, it would be very hard to be sure that this could be done safely. Would an embryo formed from such a manufactured egg or sperm grow into a healthy baby? Who knows? There's no way of testing it without trying, and no guarantee that if this works in animals it will be safe in humans. On the other hand, the history of this type of technology has been that if it is possible then someone, somewhere will give it a go.

Let's leave Adam and Steve now, and consider a heterosexual couple who are planning a baby. John and Jane have dreams for their yet-to-be-conceived child. They want him or her to be healthy, of course, but perhaps they hope for more than that. Jane, sporty herself, hopes for a child who might become an Olympic athlete. John, raised in a difficult home environment, wishes for a child who will be compassionate and kind. They know that being smart and attractive gives people an advantage in life, so of course they want those things for little Oscar or Sophie as well.

Can they choose? Should they be able to, if they can?

The diagnosis of genetic disease is unique in that it can be done not only before a person is born — prenatal diagnosis

— but *before an embryo is even implanted in the mother's womb*. Pre-implantation genetic testing (PGT) isn't just for mitochondrial conditions; we can test a bundle of cells too small to even see without a microscope, and say, 'If this embryo grows into a person, one day he or she will have Huntington disease' (or any of thousands of other conditions).

For PGT, in-vitro fertilisation is used to make as many embryos as possible (which is also the usual goal when IVF is done to treat infertility). The embryos are allowed to grow for a few days until they form a little ball of cells. In the most delicate of medical procedures, a few cells are sucked out, and DNA from those cells is tested for a specific genetic condition. Then, an embryo that is known not to be affected by the condition is implanted. This way, the parents can be sure from the beginning that their child will not be affected.

We can do PGT for single-gene conditions only if the exact genetic cause of a condition in a family is known. Even so, it's pretty common for the idea of 'designer babies' to be raised when PGT is discussed. This is obvious nonsense in relation to the way the testing is actually used in most countries, but it's easy to understand where the idea comes from. If you can choose to avoid a bad outcome such as a fatal genetic disease, could you select in favour of something you think is desirable in your child, such as intelligence, height, beauty, or sporting ability?

The closest this gets to a 'yes' is using the technology to choose to have a boy or a girl. In some countries, boys are favoured over girls, to the extent that those who can afford it sometimes use PGT to choose to have a boy. It seems likely this has been widely done across the world. In some countries, the practice is banned — even those, like Australia, where there has not been any particular bias towards one sex or the other among parents who seek selection. Using PGT for 'family balancing' is frowned on, and even banned, for reasons I have never understood. If you have two boys and would like your third child to be a girl, there would be no obvious harm to that girl or to her brothers, and no apparent damage to society at large from that decision. Would she be a 'designer baby'? Only by the very broadest of definitions. As we shall see in chapter 9, the way that characteristics like height and intelligence are inherited makes it unlikely that we will be able to pick the 'best' embryo from those that are available. Even if we could, we wouldn't have 'designed' the baby, we would just have selected one who could have been naturally conceived anyway.

So — if you can't *choose* your preferred embryo in any meaningful way, can you *manufacture* an embryo to suit? The answer is, a very qualified, yes. The genie is out of the bottle, but this genie is not necessarily to be trusted.

Genie: Okay, Dave, you get one wish. Use it wisely!
Dave: I wish I was rich.
Genie: You got it! Enjoy!
Rich: Hey, wait a minute …

It has long been possible to modify the genes of a living creature. We talk about 'making a mouse' or 'making a fly' (or a fish, or a worm) in order to study what happens when the function of a particular gene is altered. This can include knocking a gene out, so it doesn't work at all, or adding a specific mutation to a gene, or just putting in extra copies. You can put genes from one creature into another — a famous example is putting the gene that makes some types of jellyfish fluorescent into other animals, so that you get a rabbit that glows green when you shine the right sort of light onto it. There are practical applications to the jellyfish protein, it's not just a genetic engineering party trick. For instance, if you want to work out if a specific protein is needed in a particular developing tissue, you can splice in the jellyfish protein next to the gene you're interested in, and then see which bits of an embryo glow green. The green glow marks out very precisely where the protein you're interested in can be found.

Genetically modified organisms are already economically important — mainly in crops, of which there are many, but there are also a number of animals that have been modified and will probably find their way to your plate soon, if they haven't yet. Could you modify humans? Of course you could — what works for one mammal works for another. But what type of genetically modified human (GMH) would you make?

One option might be to create a GMH with big muscles. There's a gene for a protein called myostatin, which acts as an off-switch to muscle growth. Some people aspire to look like Arnold Schwarzenegger in the 1970s, but, from an evolutionary perspective, there's an advantage to keeping the growth of muscles in check. Making muscles takes nutrients, and then once you have them, you need extra food to keep them going. Limiting muscle growth is all about resource management: efficiency demands that we have enough muscle, but not too much. But if resources are no obstacle ... well, there's a breed of cow called the Belgian Blue that has *enormous* muscles because it has faulty myostatin. That's an advantage (although not to the cow) if your plan is to eat that muscle. Would it be helpful otherwise?

There's at least one human being reported who lacks myostatin. When last described, 14 years ago, he was an extraordinarily heavily muscled and strong four-year-old. Presumably, he is now an extraordinarily heavily muscled and strong young man. So, you might say — what's the problem? Let's get cracking and create an army of Olympic weightlifters!

There are two problems, and both relate to safety. When you modify an organism's genes, there's a chance that things might go wrong. In the process of knocking out myostatin, you might inadvertently take out something else you would rather have left alone. You could wind up a with a child who is very muscular, but who has a high risk of developing bowel cancer at a young age, or who is born with severe epilepsy. The second problem is that even if everything goes perfectly, we don't know what the long-term effects might be for people with no myostatin. Perhaps they will remain healthy all their lives. But one healthy four-year-old gives you no grounds for confidence about that. Belgian Blue cattle are somewhat fragile — they struggle in harsh climates, they have difficulty giving birth, and their fertility is lower than that of other breeds. It may be that some of those problems are separate from their enhanced muscles, but there is no way of knowing if a GMH lacking in myostatin might also have long-term health problems.

Or maybe you are after basketballers. Growth hormone, as its name implies, makes you grow. So how about a GMH who has extra growth hormone? Well, we already know how this one turns out. It would indeed produce people who are tall — very tall — but they would very definitely have other health problems as a result. We know this because there is a naturally occurring version of this GMH as well. André Roussimoff, known as André the Giant (who so superbly played Fezzik in *The Princess Bride*) had this condition: acromegaly. His over-production of growth hormone was caused by a tumour, rather than by any genetic difference, but the effect would be the same either way. Indeed, Roussimoff was tall (224 cm/7 ft 4 in.), and strong, a professional wrestler. But with his size and strength came overgrowth of his facial features and a host of health problems, leading to his death aged just 46.

In late 2018, it was announced that a Chinese scientist called He Jiankui had used the CRISPR technology, a powerful tool for editing genetic material, to change the genetic make-up of two babies in an attempt to make them resistant to infection by HIV. Later, it was revealed that a third baby had also been born. Dr He's stated goal was to introduce a specific change in the gene *CCR5*. This gene codes for a protein that sits on the surface of some white blood cells, and is exploited by HIV to get into and infect those cells. The particular change Dr He was trying to introduce is fairly common in Europeans but absent from Asians; people with two copies of the changed protein (about 1 per cent of Europeans) are resistant to HIV infection.

The couples recruited for He's research were chosen because, in each case, the father had HIV. On the face of it, this makes He's intentions seem like they might be justified: he wanted to make babies who would not be affected by their father's HIV;

a purely medical motivation. Except that this would have been medical nonsense: there are already sperm-washing techniques that mean that it's possible for an HIV-positive man to father a child with an extremely low risk of being born infected by the virus. There's no medical justification for changing such a child's DNA to further reduce that small risk. Dr He himself, interviewed in November 2018, stated that his goal was to protect the babies against HIV infection *later* in life. So what was the point of recruiting HIV-positive fathers? Your guess is as good as mine. He doesn't seem to have offered any coherent explanation.

Later, it emerged that although He did succeed in modifying the *CCR5* gene in the babies, he didn't manage to introduce the specific change found in Europeans, meaning that there's no way of knowing for sure whether those will be resistant to HIV — particularly because it seems that the changes only affected some of the babies' cells, i.e. they are mosaic. There are potential downsides to having *CCR5* that doesn't work normally, with vulnerability to some other viruses being part of the price paid for HIV resistance. It's not at all clear whether these three babies have been helped or harmed, and that's assuming that the gene that was targeted is the only gene that was changed in the process.

Sound ethically dubious to you? It did to the Chinese authorities, too, and an investigation revealed that He had acted without proper ethical approvals in place. In 2019, he was jailed for three years and heavily fined for his actions.

This is not to say that there are no conceivable medical reasons why you might edit an embryo's DNA. Almost always, when there is a specific genetic condition in a family, it's possible to use PGT to select embryos that are not affected by that condition. Sometimes, though, there are circumstances in which this may not be an option. Suppose, for example, that both parents are affected by the same autosomal recessive condition (perhaps they met in the waiting area for a specialised clinic). Both parents have two faulty copies of the gene in question; neither has a normal copy. This means that every embryo they conceive will also be affected, so there would be no unaffected embryos for PGT to choose. At present, if they want to have a child who is not affected by the same condition, they would have to consider adoption, or the use of donor eggs or donor sperm. Gene editing could open the door for them to have healthy children who are biologically their own. Would that be such a bad thing? It seems possible that with careful preparatory work, likely to take years, this may one day become a standard medical procedure.

Having said that, many would think that safety concerns are not the only objections that should be raised to deliberately changing the genetic make-up of a human being, particularly in a way that can be passed on to that person's own children. Some would argue that this is playing God, against nature, or otherwise an ethical minefield. Whether you accept that or not, there can be no doubt that deliberately modifying an embryo to create an 'improved' human raises extra questions about safety, not only from the procedure but from the changes that you are aiming to make, that seem impossible to answer. Only someone completely unscrupulous would attempt such a thing.

Which means, of course, that, by now, someone, somewhere, has done it already. There is surely a genetically modified super-baby out there in the world. For her sake, and for the sake of her future cousins, I hope my concerns about safety are wrong.

9

Complexity

At the dawn of the twentieth century, it was already
clear that, chemically speaking, you and I are not
much different from cans of soup. And yet we can
do many complex and even fun things we do not
usually see cans of soup doing.

PHILIP NELSON

Speaking of safety … thalidomide is a remarkably useful drug.
By suppressing the growth of blood vessels, it effectively treats
a common and serious complication of leprosy, and it is active
against a type of cancer called multiple myeloma, as well as
against certain inflammatory conditions. Unfortunately — as it
turns out — it's also a really good treatment for nausea, and
works as a sleeping pill without causing addiction. The latter
was a powerful claim when the drug was introduced, because it
was a time when barbiturates were widely used and were known
to cause death from accidental overdose,[1] as well as being highly addictive. Unlike barbiturates, you can swallow a handful
of thalidomide tablets and survive the experience without any
immediate harm — a handy angle for marketing purposes.

---

1   Also deliberate overdose — Marilyn Monroe was only one of many who
     died this way.

Advertising for the drug highlighted its safety: an advertisement in the *British Medical Journal* in 1961 proclaimed the drug 'highly effective ... and outstandingly safe'. One advertisement showed a toddler holding an open medicine bottle, with the message being that, if the bottle had contained barbiturates, the toddler would be in mortal danger. Luckily, the bottle was instead full of nice, safe thalidomide.[2]

In fact, the drug had hardly been tested for safety at all. A few years after its introduction, it emerged that thalidomide could cause damage to nerves, resulting in chronic limb pain. Then, in late 1961, the first reports of damage to unborn babies started to trickle in. The drug was withdrawn from sale almost immediately in most countries.

Sadly, this was too late to prevent an unprecedented medical catastrophe. Thalidomide's anti-nausea effects had led to it being marketed as a treatment for morning sickness during pregnancy, starting in the late 1950s. As a result, thousands of children were born with physical malformations. The most striking of these were severe limb deficiencies — being born with missing hands or arms, or with all four limbs reduced to small remnants or being entirely absent. Other problems, such as congenital heart disease and kidney malformations, were also common. Many affected babies did not survive infancy, and many pregnancies were miscarried due to the effects of the drug.

Among developed nations, the United States stood apart in being almost completely spared the ravages of thalidomide. This was thanks to the brilliance of a doctor who was working for the Food and Drug Administration (FDA) at the time of the drug's introduction elsewhere in the world. Dr Frances

---

2   The modern childproof lid for medicine bottles was invented in 1967
    by Mr Peter Hedgewick, following efforts over a five-year period to
    encourage development of such a lid by a Canadian paediatrician, Dr
    Henri J. Breault. Breault was sick of treating children who had accidentally
    overdosed and was determined that such overdoses should be prevented.
    The idea caught on quickly in Ontario, but, despite the obvious benefits
    of the invention, it took some years for its use to become widespread
    elsewhere.

Kelsey reviewed the safety information provided by the drug's manufacturer. She found the evidence unconvincing. Six times applications were brought to the FDA, and six times Dr Kelsey rejected them — saving unknown thousands of children.

Kelsey was a remarkable woman. Born in Canada[3] in 1914, she trained in pharmacology at McGill University, receiving her Master of Science degree in 1935. She applied to the newly established pharmacology department at the University of Chicago for a research-assistant position, which came with a scholarship that would allow her to study towards a PhD. She was delighted to receive a letter of offer, but concerned that it came addressed to *Mr* Kelsey; it appears her name had been confused with 'Francis'. She asked her professor at McGill if she should send a telegram explaining the point. He told her that would be ridiculous, and advised her to reply accepting the position, but writing (Miss) after her name. Interviewed by the *FDA Consumer* magazine in 2001,[4] Kelsey said, 'To this day, I do not know if my name had been Elizabeth or Mary Jane, whether I would have had that first big step up. And to his dying day, Professor Geiling [her supervisor in Chicago] would never admit one way or the other.'

Kelsey's first big impact on medicine came while she was studying with Geiling. He was asked by the FDA to investigate a series of deaths in people treated with a new antibiotic, sulphanilamide. Kelsey helped to discover that the method used to prepare a syrup version of the drug (which had an unpleasant taste in pill form) involved the use of a poisonous but sweet-tasting solvent, diethylene glycol.[5] The syrup tasted good, cured infections, sold very fast ... and, because it had not been tested for safety, also killed 107 people, many of them children. This disaster seems to have left a deep impression on

---

3   Canadians seem to have made a disproportionate contribution to the cause of drug safety.

4   Like you, I always keep a stack of these on my bedside table.

5   A similar compound, ethylene glycol, is used as antifreeze in car radiators, and is also poisonous.

Kelsey, and doubtless shaped her thinking in relation to drug testing. As you might expect, it had a similar effect on the country as a whole, and led to legislation, the Federal Food, Drug, and Cosmetic Act of 1938. This law has had a profound and long-lasting impact on the way that the safety of medications is managed. The Act required drug companies to show evidence that medications were safe before they could be marketed, and to provide warnings of potential hazards. This seems an obvious idea now, but it was a major shift from the free-for-all that had prevailed beforehand, and, because of the importance of the US as a centre for drug development and as a market, has had a major and lasting effect on the management of drug safety in the rest of the world as well.

In 1960, Kelsey was offered a job at the FDA. The very first task she was given was to review the application for approval of thalidomide. The drug had been used in Germany since 1957 and had rapidly spread through the world after that; it was generally believed to be safe. Kelsey was given the job of reviewing this particular application because it was thought a good idea to give new employees a relatively easy assignment to start with(!). Fortunately for thousands of American children, Kelsey's training, experience, rigorous approach to the evidence presented, and refusal to be intimidated by the drug company (which put enormous pressure on her to relent), all made her ideally suited to assess this particular application.

For once, the rest of this story is not that of an unsung hero whose work was disregarded during their lifetime. Kelsey received the credit she richly deserved. *The Washington Post* ran a front-page story entitled '"Heroine" of FDA Keeps Bad Drug Off of Market'. In 1962, President John F. Kennedy presented her with the President's Award for Distinguished Federal Civilian Service, the highest honour the US government gives to its civilian employees. In 2000, she was inducted into the National Women's Hall of Fame. She retired in 2005, aged 90. In 2010, she received the FDA's inaugural Drug Safety Excellence Award, which was named after her and is given annually to an FDA employee. Kelsey died in Canada in 2015, aged 101, shortly after being presented the Order of Canada by the Canadian governor-general.

By contrast ... I know this is probably unfair on her, given that I am writing from a position of hindsight, but I can't resist telling you about a letter written to the *British Medical Journal* by one Paula H. Gosling, published on 16 December 1961. It is a nice example of a longstanding tradition of medical resistance to change. Gosling roundly criticises the decision to withdraw the drug from sale. 'I must protest against the action of the Distillers Company in withdrawing ... thalidomide ... from the market', she says, and then argues that, if this had been done because of the reports of nerve damage, that would have been one thing, but 'to do so on the grounds of two unconfirmed reports from alien sources of a possible — not proved — association of thalidomide given in early pregnancy is quite irresponsible'. Charmingly, she goes on to point out that because thalidomide is tasteless and odourless, it is the ideal sedative for children ... and cats! Lastly, she suggests that 'any doctor alarmed by these unconfirmed reports' could prescribe other drugs 'for pregnant females' ... and concludes, 'Have we lost *all* sense of proportion?'

No doubt Gosling came to regret this letter before much time had passed.[6]

If a drug that is taken during pregnancy causes malformations

---

6 In fairness, I should confess that I have done something rather similar. I wrote a testy letter to the journal *Nature*, following the early reports that babies whose mothers were infected by zika virus during pregnancy could suffer damage to their brains, causing them to be born with small heads and go on to have neurological problems. The letter mainly complained about sloppy use of terminology, but also pointed out that, so far, there wasn't really a lot of hard evidence for harmful effects from zika infection. Within a week of its publication, a paper came out in *The New England Journal of Medicine* that nailed down the relationship, beyond any reasonable doubt. I had vigorously and thoroughly plastered my face with egg, while standing on the most visible platform in science.

in the baby, it is called a teratogen.[7] Thalidomide is one of the most potent teratogens known — even a single tablet, taken at the wrong time, can have devastating effects. And yet … and yet there were babies whose mothers took thalidomide during the critical time in early pregnancy, when the fetus is most sensitive to its ill effects, and were nonetheless born whole, hale and hearty, somehow unharmed. Very likely, the babies born un-scathed by their exposure to thalidomide were a small minority — but there is good evidence that they exist. How could this be?

A few times a year, I give a lecture to medical students about genetics. The idea is to give them an overview of the role of genetics in medicine, reminding them of the fundamental princi-ples they were meant to have learned earlier in the course, and trying to give them a feel for where the field is going. If you're a medical student today, it's pretty much guaranteed that some of your future patients will be having genetic tests, and, if you're going to order a test and receive a report, it helps to have some idea of what the results might mean. Anne Turner used to give the same lecture, which she called 'genetics in an hour', and, in the years since I took it over, I find myself constantly realising that some new development should be added, then discovering that the lecture is too long and deleting slides.

Even so, I always take time to make the case that each and every other branch of medicine should be viewed as just a sub-specialty of genetics. Pretty much *everything* that afflicts human beings, and everything about us that is not an affliction, too, has genetics at its core.

Take trauma, for instance. You may not think that being involved in a car accident, or getting punched, is a genetic problem. Consider, though, that there is a major genetic risk

---

7   The word derives, rather unfortunately, from a Greek word meaning 'monster'. Medications are only one type of teratogen — other drugs, such as alcohol, and maternal infections, such as zika, can also be teratogens.

factor that strongly influences whether these things will happen to you. It is called the Y chromosome. At every age after 12 months — the age at which most of us are able to get up, about, and into mischief — males are more likely to be injured than females.

You may have some thoughts about why this should be the case. The obvious culprit is testosterone: aggressiveness and impulsiveness are characteristics that will put you in harm's way, and testosterone is known to promote both.

That may well be a contributing factor, but most likely it's not as simple as that — there are probably other ways that the Y chromosome influences behaviour. Of course (despite my desire to put genetics at the centre of the medical universe), we should not imagine that the differences in behaviour between men and women are due only to their different and inborn physical and chemical make-up. There are social as well as genetic influences on maleness and the way that men behave. If you are conditioned from birth to think that you should be adventurous and bold, nobody should be very surprised if you do wind up adventurous and bold. If you're taught that your place is in the home, and that quiet, passive pursuits will suit you best — well, you might cut loose, but, on the whole, there's a decent chance that this upbringing will affect your choices in life, in ways that tend to keep you out of harm's way.

Yet to muddy the waters still more, there are other genetic influences on your chance of getting hurt, whether or not you are encumbered by a Y chromosome. Not all men have the same risk of getting hurt as each other, and there is plenty of overlap in behaviour (and risk) between men and women. Both men and women can be more or less impulsive; some men (and women) are more likely to stay home and play video games, while others prefer to go out BASE-jumping. Those differences between members of the same sex are also under genetic control to some degree, but understanding them is not as simple as being able to point at a single errant chromosome and blame that.

In short, the genetics of trauma are complex, and there is an interplay between genes and environment. That balance is not always the same, and there are times when environment can completely overrule genetics. Put the most peaceable, unadventurous woman in Baghdad and she may, by sheer bad luck, be at the market when a car bomb explodes. Put the most (potentially) reckless and aggressive of men in an environment where men are socialised against violence and other opportunities for harm are limited (you'll note that I couldn't think of a good example of such an environment, but all things are relative) and he may live, unscathed by any violence, to a ripe old age and then die in his sleep.

This type of interaction between genes and environment has an effect that is easiest to study and understand in a population, rather than at the level of a single individual. It's a bit like the difference between climate and weather: we know that, on average, men are more likely to experience violence, just as we know that, on average, it's hotter in summer than in autumn. Nonetheless, it's not terribly surprising for there to be a cool day in summer or a warm day in autumn; the individual days are like individual people. Just as we can't predict what the weather will be like a year from today, even perfect knowledge of a baby's genetic make-up will never tell us exactly what type of person they will become, or even exactly what kind of health problems they may experience.

Virtually all human diseases have some contribution from genes. This ranges from conditions like those that mostly fill this book — in which a single gene is faulty or there is a problem at the level of a chromosome, and that is enough on its own to cause a genetic condition — through to common conditions like stroke, in which genes and environment are both important, and the genetic contribution is not from a single gene but from many. That latter scenario involves many, many different genetic elements, each of which gives a tiny nudge to the chance that you will have a stroke. Some nudge your risk up, some nudge it down. It's likely that there are hundreds or even thousands of such genetic nudges acting on the risk for each common medical condition; for most people, most of the time, any one of these only has a very small effect. A genetic variation that increased or decreased your chance of having a stroke by 20 per cent would be considered to have a very strong effect. A catalogue of genetic influences on stroke[8] identified 287 different places where there was good evidence for such a nudge, most of which only have only a slight impact on risk.

The main way that we have identified genetic influences on common disease, so far at least, is through genome-wide association studies (GWAS). If you want to run a GWAS,[9] you need very large numbers of people — tens to hundreds of thousands[10] — about whom you know something. Perhaps you know how tall they are, or what their blood pressure is, or whether they have had a stroke. You get DNA from each of them and look at thousands of places, spread across the genome, where there is known variation between people. Then you compare these genetic results with the known information about the people in your study. The aim is to find a link between the DNA results and the characteristic you are studying.

Suppose there is a particular spot where some people have a C whereas others have a T. When we look at people who have *not* had a stroke, we find that 50 per cent have a C and 50 per cent have a T. Then we look at people who *have* had a stroke, and find that 60 per cent have a C and 40 per cent have a T. People with stroke are more likely to have a C than the controls.[11] The next step is to do the study all over

---

8   Part of a catalogue kept by the National Institutes of Health, accessible at https://www.ebi.ac.uk/gwas/home

9   Personally, I have no desire to do this, but I won't judge you if you decide to give it a go.

10   Putting together data from such large numbers of people often requires huge collaborations. In 2014, a study of the genetics of height published in the journal *Nature Genetics* had 445 authors, as well as four groups who were not individually named. Yes, I counted them.

11   If the numbers had been the other way round, C would have been *protective* against stroke rather than a risk factor.

again, in a 'replication cohort' — a second group of people — to demonstrate that the link you found the first time around wasn't a fluke. This is necessary because, in the early days of GWAS, there were numerous GWAS 'hits' that turned out to be statistical blips with no relationship to reality. The statistical bar for that second study is a bit lower than for the first, because you're looking at one particular target rather than scanning the whole genome. This means you don't need quite as many people in your second group, but it's still a lot of work. Let's say that, second time around, you find something similar to your original result. Congratulations, you've found a risk factor for stroke!

... but it may not help you all that much. For a start, it may well be that this genetic difference does not, of itself, have any direct bearing on the risk of stroke. It could be an innocent by-stander that just happens to be sitting, minding its own business, somewhere close to some other change that is the actual culprit. This means that identifying a variant like this is often only the first part of a long and frustrating search for the actual villain in the piece. The second problem is that the information is pretty meaningless for any one person. If half of the population have a C, and there's only a small extra risk of stroke in people who have a C, you shouldn't get too worried if you find that you have it, too. And the information may not be meaningful for people from a different population — the link between this particular C and stroke might only hold true in people from a European background, for example. There is an unfortunate oversupply of studies done in people with European ancestry, unfortunate because of the serious lack of similar studies in people from other populations.

There is a pretty good chance that your GWAS hit does not sit inside an actual gene. Most do not. Sometimes that's because of the innocent-bystander phenomenon mentioned above — there *is* a change in a gene that's important, and the C that you found is sitting *near* that change. More often, though, if we can find a specific reason why a particular variant is associated with a condition we're interested in, it relates to how genes are controlled, rather than to changes that result in a different version of a protein being produced. The genome is full of sequences that are important for regulating activity in the cell nucleus. This often works by producing signals in the form of RNA, a chemical that is nearly-but-not-quite the same as DNA. *This* activates *that*, which suppresses *the other thing*, which in turn changes the activity of a gene, meaning more or less of a protein being produced ... which might be relevant to the thing you were interested in to begin with. Our understanding of this network of signals is far from complete, but it's likely that many GWAS hits involve subtle shifts of balance in a tangled web of information — not something we are likely to figure out in a hurry.

Ideally, we would like to identify all of the genetic variation that affects human[12] diseases, as well as characteristics like height. If we could completely understand the genetic influences that decide whether a particular person will have a heart attack, it might be possible to find new ways of preventing that from happening.

Long before the human genome was sequenced, there were efforts to figure out how important genes are in controlling various conditions and characteristics. Think 'nature vs nurture', with an effort to actually measure the 'nature' part. A widely used measure in this area is heritability — the degree to which

---

12   My focus in this book is on humans, but the same technology is widely used in other organisms, including for the purposes of agriculture. If you can identify genetic variation that influences milk production in cows, the dairy industry will beat a path to your door. There are plenty of similar applications in other areas, from crop production to horse racing — and it goes beyond the more obvious organisms. For some years, until he retired, I collaborated with Professor Chris Moran at the University of Sydney. Chris was part of the team that sequenced the genome of the saltwater crocodile, and he worked on finding variation that affected the speed at which young crocodiles grow, as well as characteristics of their skin that are important in crocodile-leather production. Virtually every economically important organism — from honey bees to rice to farmed salmon — has had its genome studied in an effort to figure out how to make them more productive, and lucrative.

variation in a particular characteristic within a population is due to genes rather than environment. The name makes it sound like this is a direct measure of the 'nature' part of the equation, but that's not quite right: it really is about *variability* within the population. To illustrate this point, imagine you are studying hair colour in two different groups. One group is composed entirely of Nigerians; the other is composed of a random sampling of Brazilians. The Nigerians all have black hair, so there is no variation to measure: heritability will be zero. That doesn't mean genes aren't important in controlling the hair colour of Nigerians — quite the opposite is true. The Brazilians have everything from black hair to blonde. This variability is mostly explained by genetics, so heritability will be high.

There are several ways of calculating heritability. A common and relatively simple approach is to compare the degree to which identical and non-identical twins are similar. The assumption is that twins share their environment equally, even down to the conditions they experience while still in the womb, and that environmental effects are no different for identical and non-identical twins. Since identical twins share all of their genes,[13] whereas non-identical twins share, on average, only half of their genes, you expect (and usually observe) that identical twins will be more similar to each other than non-identical twins — not just in physical appearance but also in height, blood pressure, and so on. Some fairly simple maths, by the standard of this sort of thing, lets you measure that difference across a group of twins, and use it to calculate heritability. Heritability scores, however calculated, range from zero (no contribution from genes to the variation in that population) to one (the variation is entirely due to genetics); they can also be expressed as percentages.

---

13   In principle. In fact, it is possible for identical twins to have differences either in the sequence of their genes (that happened after the split, and that would usually be mosaic) or in some of the settings within the cells that control the functioning of genes. Usually, these are too subtle to cause differences that you can observe just by looking at them, but they are real nonetheless.

Heritability estimates for different characteristics vary between studies, and between populations. One measure that is relatively consistent across a number of studies has been height: in well-fed populations, height has a heritability of 0.8. *Most* of the variation in height is due to genetic effects. Studies in China have found lower heritability for height, around 0.65. That's still an important contribution from genes, but suggests that environment might be more important in China than in, say, the US. A possible explanation for this relates to the fact that, if your mother is undernourished while you are in the womb, and you are undernourished as a child, you are likely to wind up shorter than you would have done otherwise. Study people who have grown up in hard times and you will see a larger impact from the environment, pushing the contribution from genes down.

GWAS became possible in the mid-2000s, and really took off in about 2007. Pretty quickly, the GWAS-ologists became aware of an awkward problem: *missing* heritability. By 2010, after quite a bit of effort, GWAS studies had managed to explain only 5 per cent of the variation in height — quite a gap from the 80 per cent predicted by measures of heritability! Gradually, over the past decade and a bit, that gap has narrowed — but there is still a gap and it's not fully explained. The biggest study of the genetics of height so far used data from nearly half a million people from the United Kingdom, volunteers who had donated DNA and extensive personal and medical information as part of the UK Biobank project. A group led by Stephen Hsu (our second Hsu — of whom more later) was able to use this treasure trove of data to explain 40 per cent of the variation in height — an impressive feat, but far from allowing us to predict someone's height accurately from their DNA alone. The predictive score they developed allowed them to estimate the heights of most of a group of people (not part of the original data set) to within a few centimetres. That sounds impressive, but the range of possible actual heights for any given predicted height was pretty wide. Imagine a witness in a court case saying, 'I think the man was about five foot eight, give or take an inch or two. But he could have been as short as five foot two or as tall as six foot two.'[14] Hsu's group also convincingly showed that, for height, studying more people or looking at more genetic markers would be unlikely to provide more accuracy — they have taken this approach about as far as it is ever likely to go. They also looked at educational attainment, measured on a six-point scale that tops out at 'university degree', and found they could explain just 9 per cent of the variation in that measure, but also that an even more enormous study would have a decent chance of explaining more of the variation.

Why is it that even a superb study, using huge amounts of data, can only explain part of the variation that should be there to find? There are two main explanations that are put forward. The first argues that the traditional measures of heritability are substantial overestimates — for example, a recent study using data from Iceland and incorporating genetic data (from whole genome sequencing) in calculating heritability found rather lower values than from the traditional approaches. For height, rather than heritability of 0.8, the new estimate was only 0.55 — suggesting that Hsu's group found more than 70 per cent of all that there was to find. For body mass index (BMI),[15] the numbers were 0.65 (old method) and 0.29 (new method); for educational attainment, it was 0.43 and 0.17 — both very substantial differences.

Perhaps more interestingly than just a miscalculation, an alternate explanation for missing heritability is that there are huge numbers of undiscovered genetic contributors to the characteristics that are being studied — but most of them only have very, very tiny effects, that are too hard to measure even if very large numbers of people are studied. This idea is not

---

14   Or, if you prefer metric, 'I think the man was about 173 cm tall, give or take a few centimetres, but he could have been as short as 157 cm or as tall as 188 cm.'

15   A measure of the relationship between height and weight; it is weight in kilograms, divided by height in metres squared.

new — its centenary was in 2018. In 1918, R.A. Fisher, one of the luminaries of modern statistics, proposed the infinitesimal model, in which a variable characteristic like height is under the control of an infinitely large number of genes, each of which has an infinitely small impact on the characteristic, as well as, of course, environmental influences. Fisher wasn't suggesting there really are an infinite number of genes, it was just a way of thinking about the problem.

It's possible you haven't heard of Fisher, but in certain circles he remains a revered figure. This is not so much for the infinitesimal model — a relatively minor achievement by his standards — but for many other major contributions to statistical and genetic theory. Ian Martin, the scientist who taught me how to breed mice,[16] retired a few years ago. In December 2016, the University of Sydney honoured Ian's contributions to the field of mouse genetics, and to the university, by awarding him an honorary doctorate in veterinary science. At a reception held before the ceremony, Ian told me in detail about the time that he met Fisher and showed him some of his work, which Fisher commented on favourably. Ian was awarded his PhD in 1962, the year that Fisher died, so the events must have occurred well over half a century before, but it was obvious that their encounter was as clear in Ian's mind as though it had happened the previous week. I was deeply impressed to hear that my friend and teacher had known the great man.[17]

Let us drag ourselves reluctantly from scientific hero-worship to look at 'environment' a bit more. If the potential

---

16   For scientific purposes, not as a hobby.

17   I wish I could say that Fisher's reputation is as a giant of statistics and nothing else. Unfortunately, his reputation is somewhat blemished by his membership of the Eugenics Society at Cambridge, and by his firmly held views that there are real and important differences between the races of humans. A small positive of this, for me at least, is that through Fisher I am connected in a surprisingly small number of steps to Charles Darwin. I know Ian Martin, who met Fisher, who knew Horace Darwin — a fellow member of the Eugenics Society, and son of Charles Darwin. That's just four degrees of separation. From Darwin!

environmental influence on your health is a speeding truck, it's fairly obvious. If it's an infectious disease, you might be tempted to think not only that the environmental component (the virus, bacterium, fungus, or parasite) is obvious, but that it's the *only* thing that matters. If so, you'd be wrong. Some people have immune systems that deal well (or poorly) with particular types of infection. At one extreme, there are people who are almost immune to infection by HIV, as we've seen (in chapter 8). At the other are people with genetic immune deficiencies, whose bodies are terribly vulnerable to some types of infection, or to any infection at all. All the rest of us lie somewhere in between those extremes, along a spectrum of susceptibility.

There's an interaction between host and infection, and not all germs are created equal, so that both sides of this equation are variable. Someone who is naturally resistant to influenza encounters a mild strain of the flu and doesn't even notice that they are ill. Someone else, with an average immune system, encounters a flu virus that is average in nastiness, and has a miserable few days but recovers with no lasting effects. And someone with a vulnerable immune system has the bad luck to run into a particularly gnarly virus ... like the H1N1 strain that rampaged across the world in 1918 (Spanish flu) and 2009 (swine flu) ... and that person dies.

In genetics, a particular focus of our thinking about the interaction between genes and environment is the early months of pregnancy. Thalidomide is an extreme and infamous example of an environmental influence on early development, but there are other medicines that are known to have the potential to harm the unborn baby. Isotretinoin, marketed as Accutane or Roaccutane, is very effective at treating acne. Unfortunately, if a woman becomes pregnant while taking this drug, there is a very high risk of a variety of problems in the baby — physical malformations and intellectual disability — although many babies are unharmed. Some drugs that are used to treat epilepsy can cause problems for the baby as well, which can lead to a difficult choice in women who have severe epilepsy and need to take a drug that is risky to use in pregnancy. Do you accept the risk to the child or switch drugs, knowing the mother might have seizures, which come with their own problems?

It's not just medications that can potentially hurt the developing baby. Alcohol can cause significant harm, particularly to the brain. Most of the evidence for this comes from the children of mothers who have consumed large quantities of alcohol for a large part of the pregnancy, but it's not clear whether there is any truly safe amount of alcohol to consume, so the advice that's given to pregnant women is appropriately cautious. Smoking during pregnancy can cause miscarriage, or poor growth during the pregnancy, among other problems. There are infections that are serious problems — the main reason we vaccinate against rubella is to prevent infections during pregnancy, and there are plenty of other infections which are bad news for the baby (zika virus is a recent addition to the list). The babies of mothers who are diabetics are more likely to have a range of physical malformations; this isn't a problem for women with the version of diabetes that is *caused* by pregnancy (gestational diabetes), because, by the time this problem starts, the growing fetus is fully formed. Not that gestational diabetes is harmless — far from it: it carries significant risks to mother and child, but physical malformations in the baby are not among them.

All these hazards are real, but they all share the characteristic of unpredictability. You can't be sure — even with thalidomide — exactly what will happen in a pregnancy. There's also the problem that, for many medications, there just isn't much evidence about risks in pregnancy; this has led to the development of efforts to study and monitor the outcomes in babies whose mothers took any kind of medication during pregnancy.

We don't have a full understanding of why some babies are badly affected by a particular insult, while others suffer no harm, but there's good reason to think that genetic variation is involved. That could be variation that directly protects the baby — perhaps that baby's genes that give the instructions to make arms and legs, and to keep the blood flowing to the growing limbs, are particularly robust versions. Or it could be maternal genetic factors that protect the baby. Perhaps the mother has a genetic variant that means that her gut absorbs thalidomide poorly, so that levels in her blood never get high enough to matter.

On the whole, public awareness that the mother's actions can sometimes harm her baby is a good thing. It leads women to stop smoking and avoid alcohol during pregnancy, and to be careful about the medications they take. But like many things, this knowledge can be a two-edged sword. I once saw a man in his early 40s, Barry, who had intellectual disability. He could walk, and talk a little, but was dependent on his parents for most things. His younger sister, in her 30s, was pregnant and was concerned that her child might also have intellectual disability. Efforts to make a diagnosis had been unsuccessful back when Barry was a child, in the 1970s, and nobody had tried since then. After a while, the question seemed less important to those who cared for him than managing his day-to-day health problems. Finally, though, Barry's sister asked the question again. When I saw Barry, I suspected a chromosomal problem, and so it proved when I tested him — he had a chunk missing from one of his chromosomes that was undoubtedly the cause of his problems.

The news had a profound impact on his mother, a woman in her late 60s. While pregnant with Barry, her first child, she had painted the nursery; after his problems manifested, she became convinced that she had absorbed fumes from the paint that had damaged her baby's brain. She believed that Barry's problems were entirely her fault. For decades, she had carried this needless burden of guilt; it had eaten away at her for all that time. The chromosome result lifted the burden from her. Like others I've seen in this situation, her emotions were powerful and complex; relief that she was not at fault, mixed with sadness for the years she had spent believing that she was.

Asking about events during pregnancy is an important part of a clinical genetics consultation, because occasionally there can be important clues about the cause of the problem. But even when it's obvious that nothing that happened was relevant to the child's problem, I try to remember always to ask the parents if there is something that happened that they are worried about, because people don't always volunteer the information. For many years, I was part of the clinic at Sydney Children's Hospital that looks after children with clefts of their palates, lips, or both. A cleft is a split, usually caused by a failure of separate parts to join up as they should during early development. Sometimes, I would see children who had an underlying genetic condition, be that chromosomal (like velocardiofacial syndrome) or some other syndrome that was the cause of the cleft. But most children at the clinic are perfectly healthy apart from their cleft, and, thanks to the excellence of modern plastic surgery, they generally do very well once the cleft is repaired, with surprisingly subtle scars for those with cleft lips.

Their parents still want to know why this happened, and whether it was their fault. Often, the most useful thing I did in that clinic was taking explanations *away* from people, rather than the opposite. No, the stress you were under at work was not the cause of the cleft, and nor were those antibiotics you took, or the toner you got all over your hands that time that the printer at work went berserk.

Although we can't usually point to a specific cause of conditions like cleft palate or congenital heart disease, we can have a decent crack at working out how important genetic factors are in causing them. For both of these conditions, we know that there are some families in which a change in a single gene is the main cause of the problem. I've contributed in a small way[18] to identifying single-gene causes of congenital heart disease, and my friend Tony Roscioli (who is a genetic Sherlock Holmes, with an extraordinary capacity to sift through genetic data and uncover new links between genes and diseases) has found

18   As a minor player in a cardiac genetics group led by two embryologists, Sally Dunwoodie and Richard Harvey, and a cardiac surgeon, David Winlaw — powerhouses all.

several genes linked to cleft palate. But even for 'single-gene' conditions, the outcomes are unpredictable — one child in a family might be born with a heart so scrambled that it is beyond repair, whereas another, despite having the same genetic variant, might have a minor problem, or even a completely normal heart. In this setting, an important part of the environment is something unexpected: it's luck.

It may seem like a cop-out to say, 'This child had a genetic variant that might have been fatal, but was lucky and suffered little harm, whereas this other child was unlucky and died.' But there's a solid scientific basis to this idea. At the scale of a single molecule of DNA, there can be some 'wobble' in the way that proteins and DNA interact. Likewise, the regulation of gene activity is not just a matter or on or off. Sometimes the switch flicks to a higher than usual setting, and sometimes it gets stuck on low. Over a large enough number of cells, this wobble averages out so that it (mostly) hardly matters. But when we're talking about the very early development of an embryo, when tiny clumps of just a few cells are deciding their fate, that wobble can be enough to make a real difference. Like the proverbial butterfly that flaps its wings and causes a hurricane, small changes in just a few cells at just the wrong moment in early development can have a big effect down the track.

So, we can see that there's a spectrum of possibilities that encompasses all human disease. There are conditions — like being hit by a car, or suffering the effects of exposure to thalidomide — that are mostly about environment, but are also about genetics — favourable genetic variation, which might protect you from thalidomide, or unfavourable variation, which might make you vulnerable to infection. There are conditions that are mostly caused by changes in a single gene — modified by the effects of other genes and by the environment. And there's everything else in between, with a great many conditions that are the result of thousands of small genetic nudges interacting with each other, and with the environment.[19]

With this understanding comes the possibility that we might be able to use genetic markers to make powerful genetic predictions about health, or about other characteristics. Maybe a single GWAS hit by itself can't tell you much about your future wellbeing, but what about combinations of hundreds of them? Could you put the information together and make meaning out of it? The answer turns out to be: sort of. So-called 'polygenic risk scores' have been put together for various common medical problems: stroke, heart attack, diabetes, various types of cancer, and so on; essentially this is the same as the approach used by Stephen Hsu's group to try to predict human height. The best of these are starting to become useful medical tools already. For example, a polygenic risk score can be used in combination with other information to calculate a woman's chance of developing breast cancer, in a way that can potentially change the type of screening she is offered.

Other such scores have been less successful, so far at least. A group from Germany and the UK used data from the UK Biobank including 306,473 people aged 40–73 to develop a risk score for stroke. They succeeded: using their model, the people in the top third for risk had a 35 per cent higher risk for stroke than those in the bottom third. That sounds pretty good — until you consider that they also found that by asking four simple questions about lifestyle factors, they could do rather better:

---

19 Not a genetic condition, exactly, but eye and skin colour represent something of a special case, with a relatively small number of genetic variants combining to produce most of the variation. There are handy charts available online that explain that blue eyes are a recessive trait, with pictures showing brown-eyed parents having a 1 in 4 chance of having a blue-eyed child. These have the advantages of simplicity and clarity, and the disadvantage of being completely wrong. Eye colour is in fact a complex characteristic, but, unusually, looking at variation in just a few genes — especially *HERC2* and *OCA2* — can give you a lot of information about eye colour. A group of Dutch investigators, studying a majority blue-eyed population, were able to predict that a person had brown eyes with 93 per cent accuracy using just six variants in six genes. Skin colour has similar genetics, including — unsurprisingly — sharing several of the same genes that influence eye colour.

people who had zero or one healthy lifestyle factors[20] had a 66 per cent higher risk of stroke than those with three or four. In even worse news for the risk score, the lifestyle benefits applied to everyone equally, regardless of their genetic profile. Conclusion: it's best to just advise everyone to lead a healthy lifestyle. The genetic test, based on superb underlying data and being scientifically as robust as you could hope for ... doesn't add a great deal to decision-making about how to live your life in order to reduce the chance of stroke.

That doesn't mean that scores like this won't be used by someone — if not for medical reasons, then to make money. There are a variety of companies that will test your DNA and give you detailed advice about lifestyle and diet based on the results. If you come from the same genetic background as the people who were studied to develop the scores (i.e. mostly European), the results might even be scientifically valid, more or less. They just aren't terribly meaningful for an individual. Before you pay for such a test, consider that there is really no prospect of receiving a report that says, 'It's fine for you to smoke, avoid exercise, and eat a diet that's full of sugar and saturated fats and low on vegetables.' A test result that says your body might not tolerate alcohol very well is not nearly as definitive as having a few drinks to see what happens[21] ... and so on.

A possible exception to the 'this stuff isn't all that useful' rule is testing to see how well your body handles medications. There are quite a few genes involved in the metabolism of drugs, and these vary between people — some of us make a version of an enzyme that works only sluggishly and struggles to keep up, whereas others have a highly efficient version that works particularly well. Knowing about these genes might be important. For example, if your body processes a drug fast, you'll probably

---

20   We're not talking about needing to be a vegan triathlete here. The factors they considered were: no current smoking, healthy diet, body mass index < 30, and moderate physical activity two or more times weekly. If you're a (current) non-smoker and not obese, your score is two already.

21   In the name of science, of course.

need a higher dose than other people, or you'll never have enough of that drug in your body to do you any good. On the other hand, if you're a slow metaboliser, the drug might build up in your body and poison you, so you need a lower dose than usual. Most of us either don't have these variations, or won't ever need the medications for which this matters — but without either having a test or suffering the consequences, there's no way of knowing.

Most of the companies that offer to test your DNA for complex conditions confine themselves to telling you your granny probably came from Eastern Europe,[22] and that you should eat more fruit and vegetables. At worst, some of them will also suggest you might benefit from specific dietary supplements ... that, by sheer coincidence, they are in a position to sell you. Mostly, though, as long as you don't take it too seriously, this kind of 'recreational genomics' is relatively harmless, and might even be helpful. If being told that you have a higher than average chance of developing type 2 diabetes leads you to eat better and do more exercise, that can only be a good thing.

There's always someone willing to take things further, of course. Stephen Hsu is one such person. Hsu started out as a theoretical physicist, and is evidently very good at it — he is Professor of Physics at Michigan State University. Unusually for a physicist, he has branched out into genetics. Hsu hasn't confined himself to studying height or heart disease — he has a keen interest in the genetics of intelligence, and, as we saw earlier in this chapter, has done some work looking for genetic variation that might predict intelligence. He is co-founder of a company called Genomic Prediction, which offers pre-implantation genetic testing with a difference.

---

22   If she's around, you could also find this out by asking her. But if there are gaps in the information available to you, this kind of testing is getting really quite good, although it's better for some populations than others. It will be no surprise to you at this point to learn that being of mainly European ancestry is an advantage, in terms of the likely detail and accuracy of the test results.

As we saw in chapters 6 and 8, PGT has so far been used for situations where the outcome is very clearly defined — mainly in selecting an embryo that does not have particular genetic conditions that are known to be in the family, or chromosomal abnormalities. Genomic Prediction offers this type of single-gene and chromosomal testing, like many other companies. Unlike anyone else, they also offer testing for polygenic risk scores. Specifically, they'll test your embryos and provide a risk score for diabetes (insulin-dependent and non-insulin-dependent), heart attack, high blood pressure and high cholesterol, various types of cancer, and short stature. There have been questions raised about how reliably they can assess risks for most of these. The published research on which the test apparently rests reports an 'area under the curve', a measure of test accuracy, for these conditions that sits in a range of 0.58–0.71. On this scale, a score of 1 would be a perfectly accurate test, and 0.5 a coin toss. A score in the range of 0.6–0.7 is definitely better than a coin toss, but not so good that it would be very surprising if an individual who was scored as being likely to have a heart attack didn't, and vice versa.

Set that technical question aside for a moment, however, and let's assume the test is very good at making these predictions. Is this useful information? If you were choosing between two embryos, equal in all other respects, but one was likely to have a heart attack some time during her life ... is that really a basis for making that choice?

To put that into some kind of context, one in three people[23] over the age of 45 has high blood pressure. One in 11 over the age of 45 has diabetes. Two in five have high cholesterol. Thyroid problems are less common — perhaps one in 200 people,

---

23   These are Australian numbers, but figures are similar in other developed countries. In the developing world, other causes of death, including malnutrition and infectious diseases, are more common, reducing the proportion of deaths from vascular disease and cancer, and nutritional differences make diabetes and high cholesterol proportionally less common.

mainly women, is on treatment for thyroid-hormone deficiency. Heart attack and stroke cause about a quarter of all deaths. Half of us will develop cancer at some time in our lives, and three in ten will die of cancer.

So, you're sitting in a clinic room and your fertility specialist has presented you with the results of the PGT on the four embryos from your latest cycle of IVF. Two of the embryos have major chromosomal abnormalities that would likely mean that those embryos would be miscarried if they were implanted. If not, and they somehow made it to full term and were born alive, they would be expected to have severe disabilities and short lives. Most people in this situation would think it best not to choose those embryos.

At the other end of the spectrum, though, the report says that one of the remaining two embryos is likely (not certain) to have high blood pressure, and the other, high cholesterol. Both of these are treatable problems that affect a large proportion of the population. Is this really useful information in considering whether to implant an embryo? Think about all the things this result *doesn't* tell you about this baby. A third of all great artists will develop high blood pressure.[24] Forty per cent of all great scientists will have high cholesterol. Thyroid disease, although fairly common, isn't in the same league as those others. But it's very easy to treat. A tablet once a day, a blood test now and then to check the treatment is working as it should. Not a huge burden.

Okay, but heart attacks are bad news, right? So maybe that would be a different kind of result?

Well … maybe. Remember that, already, there are things your potential child could do about their risk of heart attack. This genetic profile represents only one set of inputs into the overall risk, which will be modified by lifestyle — choosing not to smoke, eating well, exercising — and also medical

---

24   Perhaps I should declare at this point that I am on treatment for high blood pressure that needed a combination of two different drugs to bring it under control. I may not be an entirely neutral observer in this instance.

interventions. At present, the latter are mainly limited to keeping an eye on blood pressure and cholesterol (and other blood lipids) and using treatments to bring them into the healthy range if needed. It's quite possible that there will be other options available long before today's embryo becomes a 45-year-old with a heart attack.

Nonetheless, you might take the view that it's best for your child not to have to worry about all this stuff, or at least not more than most people. You might feel the same way about a risk of cancer, too. Mind you, if you add cancer to heart disease, you're talking about more than half of all deaths. If this test were a perfect predictor of possible serious health problems, you may not have that many embryos left to choose from. Is there someone in your family who has had one of these problems? Do you think it would have been better, for them and for the world, if your grandmother who had colon cancer, your father who has heart disease, or your cousin with diabetes had never been born?

Of course, there is a counter argument to this. Imagine you have two embryos and must choose one. One has a high risk of stroke, the other does not. Otherwise, the two score about the same for the various risks covered by the test. Either might turn out to be a wonderful person, gifted in some way, a boon to humanity. Or, perhaps, either might live a perfectly ordinary life but be happy, loving, and loved. Either might have a terrible health problem — say, severe psychiatric disease — not covered by the test that's available to you. With your lack of knowledge of everything else about this potential person, why *not* choose the one that is unlikely to have a stroke? Perhaps, in 65 years, that person will be starting a long and happy retirement, whereas the other would have just suffered a devastating stroke, permanently losing the ability to speak and to use one side of her body; her retirement would be spent dependent, in a nursing home. You may be long dead by then, but at least you had the chance to watch your child growing into adulthood without worrying about that particular future event. Worth it? Do you want to have this test?

It's possible to push things still further. Genomic Prediction make it very clear that they do *not* test for intelligence. But what if a similar company could tell you which embryo was likely to be the smartest? What if they could test for homosexuality? There is good evidence that, although there isn't a 'gay gene', there is a strong genetic basis to sexual orientation, and it's complex — like height, and blood cholesterol, and high blood pressure. There's every reason to think that one day it might be possible to score embryos for 'likelihood to be gay'. If you could test your embryos for this, would you? Should you?

The balance between what we *can* do in medicine, what it is *useful* to do, and what we *should* do is not an easy one to manage. There are lots of ways we get it wrong. It seems to me that even a perfect test for common, more-or-less treatable conditions is likely to do more harm than good, for most people. It also seems to me that, based on what we know about it, this particular test is too far from perfect to make it an attractive proposition just yet, even if you think the concept is good. But there is room for the test to improve … and we'd better decide what we think about it, because this particular genie is already well and truly out of its bottle.

Speaking of mythical creatures: there's magic of another kind, just around the corner.

10

A spoonful of mannose-6-phosphate

Physicians of the Utmost Fame
Were called at once; but when they came
They answered, as they took their Fees,
'There is no Cure for this Disease.'

HILAIRE BELLOC

Jesse Gelsinger died on 17 September 1999, just a few months after his 18th birthday. Gelsinger had a rare genetic condition that affected his body's ability to safely dispose of excess nitrogen. He had been a volunteer in a clinical trial that aimed to see whether a virus, which had been modified to carry a normal version of a gene called *OTC* into the cells of the liver, could safely be given to humans. Sadly, the answer was that it could not — Gelsinger's body reacted in an unexpected and extreme way. Over the course of four days, his organs progressively failed, and, despite all efforts, he could not be saved.

The clinical trial in question was one of the early attempts to perform gene therapy — a treatment designed to directly replace a gene that is faulty or missing, by inserting a working copy into the DNA of the person affected by the condition. There are formidable technical challenges that stand in the way of gene therapy, and there were some known risks from trying it. But nobody had really thought that a research subject might die in this way. Jesse Gelsinger's death was a shock to the whole field, and it seemed to many of us in genetics that the very idea of gene therapy might have died with him.

At the public announcement of the completion of the Human Genome Project in 2000, both US president Bill Clinton and British prime minister Tony Blair spoke of the ideal that our new understanding of the genome would lead directly to treatments for medical conditions. Both mentioned cancer; Blair also referred to the treatment of hereditary diseases. Over the years that have followed, there has been quite a bit of criticism of the field's failure to produce cures — a recent newspaper article asked, 'Was the Human Genome Project a Dud?', and concluded that yes, it was. Fortunately, this is far from the truth — as we've seen, the HGP has unquestionably delivered for people affected by genetic conditions and their families, in all sorts of ways. Treatments for cancers have, in fact, been developed based on knowledge gained from the HGP. By sequencing the genome of a cancer and comparing it with the genome of the same person's healthy tissues, it is increasingly possible to recognise the specific genetic damage that has driven the development of the cancer. There are many treatments — existing and in development — that work directly to counteract the effects of those changes.

But as for *cures* for genetic conditions ... well, that was always asking for a lot.

The reason it's hard to cure a genetic condition is that the problem lies so deep within. The cell nucleus is a walled fortress, protected against changes for the very reason that change is dangerous; damage to DNA can kill cells, and those that survive risk becoming cancerous. The idea of gene therapy is generally to replace something that's missing, but, for many genetic conditions, 'something missing' isn't the problem. Sometimes, the faulty gene is overactive, so the problem is too much rather than too little. Sometimes, the faulty gene causes the cell to make a toxic substance, such as an abnormal protein that builds up inside the cell and poisons it. And sometimes, even if the problem is 'something missing', it's in the form of 'something missing *at a particular time*', such as in the early weeks after conception. If that's the case, replacing the gene in an adult, or even in a newborn baby, may mean that you are acting far too late to make a difference.

Even if you can get a working copy of a faulty gene inside the nucleus, and then get it inserted into the cell's genome, and then get it to turn on and function normally — there are risks. It's hard to control where the new DNA goes; if you're unlucky, it might wind up somewhere you don't want it, and cause other problems. Most gene therapy relies on using a modified virus, since there are some types of virus that have already solved the problem of getting their DNA into the cell's nucleus — that's part of how they hijack the cell's machinery to make new viruses. But that means you have to engineer the virus in a way that stops it doing harm, without stopping it from doing the job you need it to do.

So, long before there were successful gene therapies — and as we shall see, there *are* finally some successful gene therapies — people turned to other approaches to try to treat genetic disease. For instance: if it's hard to change the DNA inside a cell, why not swap out the cell itself, replacing it with a healthy cell?

I know it's just coincidence. Still, sometimes I see several people with the same very rare condition in a short period of time, and it feels like there must be something in the water — or like the world is out to get me. In the space of a six-week period, I once saw two children whose families were presented with a deadly dilemma.

Both boys were toddlers, and their stories were strikingly similar. A series of seemingly minor problems had led them to see a range of health professionals. They had both needed hernia repairs. They had frequent colds, chest infections, and ear infections. After they turned one, they were slow to learn to walk, which eventually led to them seeing paediatricians. Each of those paediatricians ordered a screening test on urine, and that led to a devastating diagnosis.

Ethan and Angelo both had Hurler syndrome, a condition that affects the lysosome. If mitochondria are the cell's power-plants, the lysosomes are its recycling centres — structures that contain a set of enzymes that break down materials in the cell that have passed their use-by date. There are 40-odd different enzymes inside the lysosome, each with a different recycling task. One might recycle the aluminium cans of the cell, another its scrap paper, and so on. If you're born with a deficiency of the paper recycler, loads of used paper are still being delivered to the lysosome by the cell, but nothing happens to it; there's nowhere for it to go, and it can't be processed. If you put a tissue sample from someone in this situation under the electron microscope, you can see the lysosomes transformed from small balls, scattered through the cell, into large blobs that grow over time until they can fill the cell completely.

The lysosomal storage disorders manifest in different ways, depending on which enzyme is deficient. Some mainly affect the brain, others the liver and spleen; some affect nerves, or the heart. Hurler syndrome affects most parts of the body. The affected child's liver and spleen enlarge progressively. The heart valves stiffen and the heart muscle can fail. Cartilage and bone grow abnormally, so that children are very short and can have serious spine and joint problems. The tongue grows larger over time, and facial features become heavier, an appearance that is often (rather unkindly) described as 'coarse'. Joints stiffen, there can be problems breathing, especially during sleep … and, worst of all, there are progressive effects on the brain. At first, an affected child's development is normal, but it slows and then stagnates. Finally, as the neurological damage progresses, skills are lost, and, by the time of death, often before the age of ten years, there is profound disability.

This was the grim news I had to give to each child's parents about their young son.

I also gave them hope, of a sort, and a choice.

Like many genetic conditions, Hurler syndrome is part of a spectrum. With just a little bit of enzyme that works, there can be a less severe picture: the brain might be spared altogether, the first symptoms might start later, and the progression of disease might happen more slowly than in a child with the typical severe form. With just a bit more enzyme again, the picture can be very different indeed. I once met a woman in her 30s who was a little shorter than average, and who told me that she had had severe joint problems — but looking at her, I would never have guessed that she had a variant of Hurler syndrome.[1] What's more, the differences in enzyme function between those who are most severely affected and those with less-severe problems are *tiny* — one study found that children with the severe form of Hurler syndrome have about a fifth of 1 per cent of normal enzyme activity; those with the in-between form have about a third of 1 per cent; and those with the least severe form, like the woman I met, about 1–2 per cent of normal enzyme activity. It's very likely that, if you had 3 per cent of normal activity, you would be completely healthy, and virtually certain that you would with 5 per cent.

This pattern is true for many different conditions that are caused by enzyme deficiencies. The body makes a lot more of most enzymes than it needs. When researchers found this out, they drew the obvious conclusion: we ought to be able to treat people with these conditions, if we could just get a *little* enzyme

---

1   She had Scheie syndrome, which was once thought to be a separate condition but was later discovered to be effectively a less severe, or attenuated, form of Hurler syndrome. This group of conditions, the mucopolysaccharidoses (try saying that five times quickly) are thus numbered I, II, III, IV, VI, and VII. There's no V any more.

back into their bodies.

One of the ways this has been done, and done very successfully, is to manufacture enzyme outside the body and give it to people in the form of regular infusions into a vein. Getting this to work was no simple task. The first condition for which enzyme replacement therapy was developed is called Gaucher syndrome. The idea of treating Gaucher syndrome this way was suggested by Dr Roscoe Brady in the 1960s, and it took until the 1990s before there was a fully-fledged enzyme treatment on the market. Brady and his team spent much of the 1970s extracting tiny amounts of enzyme from placentas, but eventually it became possible to manufacture the enzyme using Chinese hamster ovary cells[2] that have been modified to make the human enzyme in large quantities. One of the discoveries that was made along the way was that, to get the enzyme to where it needed to be, a string of sugars attached to the surface of the protein had to finish with one particular sugar, called mannose-6-phosphate — hence the title of this chapter.

Enzyme replacement therapy works well for many storage disorders, up to a point. It's great for the soft, squishy parts of the body with a good blood supply to deliver the enzyme where it's needed — the innards, if you will. People with Gaucher disease can have hugely enlarged livers and spleens; the treatment melts those away like ice in the sun. However, for conditions where bones and joints are severely affected, the results tend to be marginal, at best. Worst of all, enzyme can't get into the brain. For someone with severe Hurler syndrome, giving them enzyme might help with some of the effects of the condition, improving quality of life for a time, but it would not change the long-term outcome.[3]

There was another option for Ethan and Angelo, a medical roll of the dice. There is a way to get healthy cells into a

---

2   I am not making this up.

3   For people with the less severe forms of Hurler syndrome, whose brains are not affected, the benefits are clearer.

person's brain, cells that are capable of producing enzyme that can then be absorbed by the brain cells. The way to do this is far from obvious: you start by poisoning the patient's bone marrow. Then you replace it, using stem cells from someone else's marrow. This is bone marrow transplantation,[4] and it was developed for treating leukaemia and lymphoma, and, less commonly, some other types of cancer. The concept is simple enough: you have a cancer of the immune system, so you get rid of all of the cells of the immune system completely, and replace them with healthy ones.[5] Another application, not surprisingly, is treating disorders of the immune system and other blood conditions. An immune system that doesn't work properly is replaced with one that will.

The reason this might work for children with conditions like Hurler syndrome is simple enough: because infection can develop anywhere in the body, white blood cells are needed everywhere. Helpfully, they produce extra enzyme, which other cells can absorb and use. Your brain contains a *lot* of white blood cells — something like a tenth of all the cells in your brain are a special type of white cell, called microglia. Replace those, and you have a rich supply of enzyme, right where you need it.

There's a catch. There are several catches. The first is that the treatment is a very big deal. The drugs that are used to kill off the child's own bone marrow are — not surprisingly — toxic. In part because of this, and in part because of the risks

---

4   More accurately: haematopoietic stem cell transplantation, or HSCT. Haematopoietic stem cells are cells that are capable of making blood cells of all types, and they include cells taken from marrow, but also can be cells taken from the placenta via the umbilical cord after a baby is born (cord blood), or even extracted directly from the blood itself.

5   There are other ways this can work. Sometimes, for cancers that don't involve the bone marrow, the patient's own marrow is extracted and stored. This allows the use of powerful treatments, such chemotherapy, that would otherwise be fatal because it destroys the bone marrow as a side effect. Once the treatment is over, the patient's marrow is restored from the backup that has been saved.

of overwhelming infection while the new immune system establishes itself, there is a substantial risk that the child might die from complications of the bone marrow transplant itself. Those who survive may have lasting side effects from the drugs, or the new immune system may attack their bodies ('graft versus host disease'). Then, even if the transplant goes perfectly, it takes at least six months for the new white cells to reach high enough levels in the brain to do any good, time in which damage is steadily getting worse. Both boys had early signs that their brains were affected already. The outcomes from a transplant for their brains were uncertain, although it was very likely they would have at least some long-term impairment. And, lastly, this would *not* be a cure — it would at best change a fatal condition into a chronic one, with gradual worsening of the bone and joint problems in particular.

Faced with a choice like this, what would you do?

In genetics, we make a virtue of non-directive counselling. The idea is that we give people information that empowers them to make their own choices, rather than telling them what to do. Not everyone I see is comfortable with this model — we are used to getting advice from our doctors, after all — and it's fairly common, in a difficult situation, for people to ask me, 'Yes, but what would *you* do?' There are good reasons why this information may not be helpful. The choice that is best for a married medical professional in his 50s, who already has adult children, may not be best for a 19-year-old single mother, or a 44-year-old couple who are facing a possible problem in what may be their only pregnancy ever. So I don't often answer this question directly; but, of course, I always have an opinion.

Not this time. I have absolutely no idea what I would do in this situation; the choice seems impossible. Accept that my child has a progressive condition that will lead to his death within the next seven or eight years, and focus on giving him as much happiness as possible in his short life? Or take a chance on a treatment that will certainly cause short-term suffering, might kill him outright, might cause serious, lasting side effects, is unlikely to completely save his brain from the effects of the disease, and will leave him with a lifetime of serious, painful bone and joint problems?

Ethan's parents made one choice, Angelo's made the other.

Whose parents made the right choice? Both. Neither. I don't think there is an objective right or wrong here.

Fortunately, dilemmas like this are rare, but it's not unusual that a treatment for a genetic condition is only partially effective, changing one condition into another, and, especially, changing a life-shortening condition into a long-term disability. We are getting better at it, and there are already many conditions for which there are specific, targeted treatments. Most of these don't work directly at the level of the gene, but tackle some aspect of the condition's biology — aiming to restore balance to some system within the cell, or in the body as a whole. Your body can't process X and it builds up, causing you harm. So we give you a medication that stops your cells from making any X, and perhaps put you on a low-X diet as well. Your body can't make Y, so we give you supplements of Y. That sort of thing.

There's a specific group of conditions, called inborn errors of metabolism — mistakes in the way the body's chemistry works — that particularly benefit from this kind of approach. For some people,[6] it can be as simple as taking a regular large dose of a specific vitamin, to kick a sluggish protein into action or prop up a chemical reaction that isn't quite working properly. By a strange twist, our patients in this group benefit from the existence of the vitamin and supplement industry. Huge numbers of people take large doses of vitamins that they do not

---

6   I don't want to give the impression that all metabolic conditions can be treated this way. Only a very small subset of people with inborn errors of metabolism have conditions that can be effectively reversed by treatment with vitamins, although there are more who benefit to a smaller degree.

need.[7] This creates a market (and competition) that results in those vitamins becoming available to the very few people who actually do need to take them, at much lower prices than would otherwise be the case. If you're someone who has been conned by the industry into taking unnecessary vitamin supplements, you can take some comfort from the knowledge that you aren't just creating expensive urine. You're also helping a group of patients with rare conditions. Thank you!

Cantú syndrome is a condition for which this type of medical balancing act really ought to work, and recently we gave it a go.

Soon after the Dutch-led discovery of the main cause of Cantú syndrome, Kathy Grange made a different type of discovery. Remarkably, on the same campus as the hospital where she worked, there was a scientist, Colin Nichols, who is internationally renowned for his work on potassium channels — the type of protein that is overactive in people with Cantú syndrome. Neither knew of the other's existence, but, the same day he read the Dutch papers and saw Kathy's name as a co-author, Colin was knocking on Kathy's office door. The pair wrote their first paper together the next year: 'KATP Channels and Cardiovascular Disease: suddenly a syndrome'. Never mind the first five words: the last three tell the story. Colin had been working in pure science, and, out of the blue, he had a brand-new human condition to sink his teeth into. He got to work immediately.

Colin is tall, lean, blue-eyed, enthusiastic. His Northern

---

7   Of course, there are people who *do* need them. If you have a vitamin deficiency that was diagnosed by an actual doctor, or have some other medical condition that requires vitamin supplements, for goodness sake keep taking them! And if you are a woman who is planning a pregnancy, it's really important that you take folic acid, because it will reduce the risk of some serious health problems in the baby, and there are some other vitamins that are worth considering. But if you started taking vitamins or other supplements without medical advice, because an advertisement told you it would 'boost vitality' or will 'make you feel better' or 'support your immune system' or something similar ... it's very likely indeed that you would be better off just eating food. Even a moderately well-balanced diet is likely to contain all the vitamins you need. Taking extra won't help you.

English accent seems a little out of place in St Louis, Missouri, where he and Kathy work, but he is very much at home there, having worked at the Washington University School of Medicine in St Louis since 1991. The first time I met Colin was in Utrecht, in the Netherlands, at the first meeting of the Cantú Syndrome Interest Group, an international group of scientists and doctors who share the goal of better understanding the condition and learning how to best to treat it. We were there for a symposium and a research clinic.

At a subsequent meeting of the group, in St Louis, Colin and his trainees presented work on mice that had been genetically engineered to have a version of Cantú syndrome. The condition isn't *exactly* the same in mice and humans — it's hard to spot when a mouse has an excess of fur, for instance — but mostly it's a pretty good match. This means you can do things such as testing possible treatments on the mice, that you might not be quite ready to try on humans. I was particularly impressed by some work on lymph vessels from the mice. People with Cantú syndrome often have a build-up of fluid in their tissues, called lymphoedema. Colin showed us pictures and videos of lymph vessels from Cantú mice, comparing them with ordinary mice. Normally, these tubes collect fluid that has leaked out from the blood into the body's tissues, and carry it back to the bloodstream. In healthy mice, the muscle in the wall of the tubes pulsed constantly, squeezing rhythmically to force the fluid towards the heart. The Cantú mice were a different story ... their lymphatic tubes sat limply, hardly moving at all. It seemed there was an obvious link with the problem of lymphoedema in humans with Cantú syndrome.

That was interesting — but what made us really sit up and pay attention was what happened when Colin's student *treated* the mouse tissues with medicine. There's a group of drugs called sulphonylureas, which work on a slightly different version of the same channel that's affected by Cantú syndrome. Instead of being found in blood vessels and lymph vessels, this channel is important in the pancreas. Yes, we're back to the pancreas, and diabetes. The reason that insulin isn't the only treatment for diabetes is that there are different reasons why people can develop diabetes. If your pancreas completely fails to make insulin, for any of a number of different reasons, then it's definitely insulin you need. When Banting and his colleagues worked out how to make a safe, reliable supply of insulin, it was this type of diabetes — called insulin-dependent or type 1 diabetes — that they were treating. In its most common form, this is an immune disease. The body's immune system mistakes the islet cells of the pancreas for a threat and destroys them.

Non-insulin-dependent or type 2 diabetes is a different kettle of fish. It's an insidious condition in which the body's cells become resistant to the action of insulin, and the pancreas gradually loses the capacity to make enough insulin to compensate. Because the islet cells keep the ability to release the insulin they do make, treatments that stimulate this release can be effective. It turns out that, if you block the pancreas's version of the channel that's important in Cantú syndrome, extra insulin is released, lowering blood sugar. Sulphonylureas work on this principle: block the channel, allowing the pancreas to release more insulin, and that insulin lowers the patient's blood sugar. Colin's student used one of these drugs, called glibenclamide, to inhibit the overactive channels in Cantú mouse lymph vessels. They sprang into life, contracting vigorously as if there were no problem.

If we have a drug that we know works to block some of the effects of Cantú syndrome in mice, and that drug is already registered for use in humans, why not just go ahead and treat all of our Cantú syndrome patients? Well, for a start, we're worried that it might be dangerous to them. Giving someone who doesn't have diabetes a blood-sugar-lowering drug might cause them to have low blood sugar, which at worst could be a life-threatening side effect. Also, there is a long history in medicine of people trying treatments that on paper ought to have worked, but which turned out in the real world to be useless, or even dangerous, for unforeseen reasons. Fools rush in where angels fear to treat. If you're going to try something like this, it has to be done in a careful, controlled way. And if you're doing that, it would be far better to find a medication that *only* works on the Cantú channel and leaves the pancreas version alone. Colin is embarking on a search for just such a drug, but such searches can take years and have no guarantee of success.

The one situation in which you *might* experiment with an untried treatment like this is one in which it seems like you have no other options.

One day in late 2017, I was called by a geneticist called Alan Ma. Alan had been asked to see a baby who had been in intensive care since his birth, several weeks before. Born prematurely, Harry weighed more than expected for a baby who had only made it to 32 weeks of pregnancy. He had a patent ductus arteriosus that had needed surgery. He had a lot of hair, including hair on his forehead that merged with his eyebrows. Alan sent me some photographs, saying, 'It looks like this boy has Cantú syndrome — what do you think?'

Alan also said that Harry had severe lung disease, and, despite all the usual treatments, he was not getting better. Worryingly, there was a report in the medical literature of an infant with Cantú syndrome who had died from similar lung disease.

It seemed that now might be the time to try glibenclamide in a human with Cantú syndrome.

Harry was in an intensive care unit, so that changes in his blood sugar, or any other — unexpected — responses to treatment could be closely monitored. The balance between risks and possible benefits seemed to favour treating. We discussed the situation with the international group, who agreed, and Alan arranged for me to meet Harry and his parents, so that I could speak with them about the idea.

Before we could try an experimental treatment on a sick baby, we had to be absolutely certain of the diagnosis. The quickest way to do this, it turned out, was to do exome sequencing. We would read the sequence of all 23,000 of Harry's genes, even though we were interested in just two: *ABCC9* and its partner *KCNJ8*. This was one of the first times an Australian lab had done this as an urgent test, and it was a complete success — in just four days, we had our answer. Harry had a change in his *ABCC9* gene that had been seen before in other Cantú children. We had our confirmation. Even better, Colin had already studied this particular change and showed that it could respond to glibenclamide in the laboratory. Alan applied to the hospital for permission to use the drug, and, cautiously, starting with a tiny dose, began the treatment.

Around the world, the Cantú community — who had agreed this was the right thing to do — held their collective breaths.

I wish I could tell you that the treatment was a miracle cure. The truth is more complex than that. Slowly, Alan increased the dose of the drug. Slowly, Harry's condition improved. His face and limbs had been swollen with fluid; the fluid cleared. His lungs gradually got better, too — not all at once, and not without setbacks, but eventually he was able to leave the intensive care unit, and finally he was discharged from hospital altogether. His blood sugar did drop a couple of times, but only mildly. We saw no evidence that the drug was harming him otherwise. Two years down the track, Harry's lungs were still working fine, and he was still taking the glibenclamide.

Did the treatment save Harry? In the paper we wrote describing all of this, we said, 'it is tempting to conclude that glibenclamide was of benefit to our patient', which I think strikes the right note of caution. I would *love* to think that the treatment helped with his recovery — and that's a problem. If you want a particular outcome, it is all too easy to fool yourself about what happened. Perhaps Harry was going to get better anyway. Maybe the drug did nothing, or actually made things slightly worse. Study a single child, and it's almost impossible to know for certain whether you have made a difference.

You might argue that it hardly matters. We treated a sick baby, his condition improved ... who cares whether the treatment made him better or he recovered on his own? The problem is that there is a particular trap that lies in wait for those who treat rare diseases. Once you use a treatment in someone who has a very rare condition, that treatment might become standard practice, even if it doesn't actually help. It then becomes next to impossible to do the studies that might show whether it really does work or not, because nobody would be willing to have their child (or themselves) potentially given a placebo instead of the treatment that 'everybody knows you use for (condition X)'.

That's not to say we should never try drugs on a one-off basis, as we did here. There are things we *can* learn from an n=1 study (a study of a single patient). We learned that the drug was not obviously harmful to Harry, within a particular range of doses. We gained some initial observations of the problems that might be helped by the treatment (the excess fluid, the problem with his lungs) and also of others that seemed not to be changed at all (the excess hair, for instance). In the ideal world, we would do a controlled study, in which patients with Cantú syndrome were treated, randomly, with either drug or a placebo, with both the doctors and the patients (and their families) unaware of which the patients were receiving. This is the randomised controlled trial, which is the gold standard when you're trying to find out if a new treatment is effective. But when the numbers are tiny and the need is great, sometimes we have to do the best we can for the child in our care; study the outcome; and report it in the medical literature so that others can learn from our experience, and add to it with theirs.

The past few years have seen an explosion of new treatments for genetic conditions, many of which are already at a very advanced stage of readiness, or have even reached the market as drugs that can be prescribed. I recently saw a French doctor, Guillaume Canaud, receive a standing ovation at a European conference,[8]

---

8   This is not normal! Polite applause is usually the best a speaker can hope for. A really good talk might be received with polite applause that goes a bit longer than usual.

after giving a talk about treatment for a complex condition in which there is massive overgrowth of limbs and tissues. Canaud had found out that the drug was being developed as a treatment for cancer, because, in some cancers, the gene involved in the syndrome (*PIK3CA*) is overactive and contributes to the growth of the cancer cells. He arranged to obtain some from the company that was working on it, and tried it on his patients. He saw dramatic improvement in some of their symptoms. In the very same week as Canaud's talk, a group led by an Australian, Ravi Savarirayan, published work showing that a new drug could lead to improved growth and other benefits in children with achondroplasia, the most common type of dwarfism. This kind of conjunction is almost the norm these days: practically every conference we go to includes news of new, targeted treatments for genetic conditions.

Perhaps most exciting of all, gene therapy is back. In one sense, it never went away. Dedicated researchers around the world have never given up on the idea of an actual cure for many different conditions. For those of us outside the field, gene therapy has had something of the feeling of a perpetually moving goalpost, always ten years away from being reality — until suddenly, in the last couple of years, gene therapies have started to hit prime time. It's a bit like the musician who is an overnight success — overnight, if you ignore the decades of practice and hard work that preceded their breakthrough album.

The basic principles of gene therapy haven't changed since the 1990s, when Jesse Gelsinger died. The idea is still to get a working copy of a gene into cells that lack that gene; viruses are still used to transport the new gene to where it needs to go. That means inside the cell nucleus, although not necessarily into the DNA of the cell — in some types of gene therapy, the working copy of the gene sits inside the nucleus and is able to function without having to change the cell's existing DNA. The decades of work it has taken to make this safe and effective have been about finding the right viruses for the job and modifying them to make them safe but still effective. The modifications stop the virus from reproducing itself, either by removing some important genes or by removing the virus's own genome altogether, so that it's only the shell of the virus that is used to deliver the replacement gene to the patient's cells.

In 2017, the Food and Drug Administration in the United States approved the first ever gene therapy for a genetic condition:[9] Luxturna treats a severe eye disease, caused by variants in a gene called *RPE65*. In 2019, the FDA approved a second: Zolgensma is used to treat a progressive neurological condition, spinal muscular atrophy (SMA). There are numerous others on the way — treatments for haemophilia, and for some immune deficiencies. It seems likely that a golden age of gene therapy is just around the corner.

But ... there's always a but. Luxturna improves vision, but it's not a cure. Zolgensma has had some dramatic successes, but it can't reverse damage that has already been done. Michelle Farrar, a paediatric neurologist, and Veronica Wiley, head of the New South Wales newborn screening lab, are leading a study in which newborn babies are screened for SMA. The idea of the study is to try to make the diagnosis before any symptoms develop, and then give gene therapy before damage to the nerve cells has occurred. The hope is that very early treatment may be able to completely cure the condition. We don't know if this will work, but the idea that it might become possible to take a baby who otherwise would have died in the first year or so of life, give them a single treatment, and *cure* them, so that they can live a completely normal life, is astonishing; it truly is the stuff of science fiction.

A note of caution is needed. Some of the babies who are identified in the first weeks of life already have symptoms. We don't know if the effects of the treatment will be permanent, or if treated children might still develop symptoms later on. And even if everything works perfectly, there is bound to be a

---

9   Gene therapy can also be used against cancer — either by directly changing the genes of the cancer cells or by modifying the patient's immune system so that it can better fight the cancer.

considerable delay before screening and treatment can become universally available, even in rich countries. We can imagine it might be like the early days of insulin treatment must have been, with a race to get enough of the treatment made and to get it to babies early enough that it will do the most good.

And then there's the cost.

The first time a patient of mine received enzyme replacement therapy, for a lysosomal storage disorder called Fabry syndrome, I held the bag of fluid containing the drug in my hand while a nurse hooked up the tubing through which we were going to administer it. My patient's mother pointed out that the contents of that small bag were worth more than her car. This was the first of what will probably be a lifetime of treatments for that boy — now a young man — treatments that are given once a fortnight. The price of a car, every other week. In the case of Zolgensma, you could substitute 'house' for 'car': the drug costs US$2.1 million for a single dose, the most expensive drug treatment ever made. At least that's a one-off — over the course of a patient's lifetime, enzyme replacement therapies work out to be *much* costlier. There are several of them, including the treatment for Hurler syndrome, that cost hundreds of thousands of dollars per year; over a lifetime, a one-off shot, even at Zolgensma prices, starts to look like a bargain. These high costs are partly because some of these drugs are just very difficult and expensive to make, and partly because the conditions they treat are so rare. If you make a successful asthma drug, it may cost a fortune to develop, but you have good hopes of selling many millions of doses. Treat a condition that affects one person in 100,000 and there's no way to spread those costs.

All of this means that society may have to face up to some difficult choices. If you have a limited health budget, how best should you spend it? Should you give enzyme replacement therapy to a child with severe Hurler syndrome, knowing that it may improve her quality of life but will not treat her brain disease, and may not greatly lengthen her life? What happens when 20 more such treatments come onto the market?

While some of the new treatments, particularly some types of gene therapy, have the potential to be a cure, it's likely that most will not. Many of the new treatments may simply turn out to change one problem into another, as with bone marrow transplants for Hurler syndrome. And there are many, many genetic conditions for which there is no realistic prospect of an even partially effective treatment.

It makes you wonder — is there another way?

II

Please, screen me

E pluribus unum (out of many, one)

ORIGINS UNCERTAIN

When Rachael and Jonathan Casella decided it was time to try for a baby, Rachael went to see her doctor, to make sure she was well prepared. The advice her doctor gave her was good, as far as it went. Look after your health; take folic acid, which reduces the baby's risk of having some serious malformations.[1] Once you're pregnant, you can choose to have a screening test to see if the baby is likely to have Down syndrome — something Rachael went on to do. Unfortunately, however, the advice didn't go far enough. Not having heard about the option of genetic carrier screening, Rachael's doctor didn't tell her about it.

Mackenzie Casella was a beautiful baby. Alert and smiley, in photographs her eyes are bright with intelligence. In a better world, she would have grown up into a beautiful young woman,

---

1 These are called neural tube defects, and they can be very severe. Among others, these conditions include anencephaly, in which the brain, top of the skull, and scalp do not form; and spina bifida, in which the spinal cord is open to the outside, and the amniotic fluid damages the cord, paralysing the baby below the level of the opening. Taking folic acid *before* pregnancy and in the early weeks of pregnancy greatly reduces the likelihood of having a neural tube defect — it's one of the great public health wins of the later 20th century.

would have gone to school and to university, and would have lived the full life her parents dreamed of for her. In the world as it is, none of those things happened, because Mackenzie had spinal muscular atrophy, and her life was just seven months long.

SMA, as you'll remember, is a progressive condition that affects the nerves in the spinal cord that control the muscles. In a person affected by SMA, those nerve cells wink out like stars in the dawn sky. Without them, messages from the brain can't get to the muscles, and the effect is muscle weakness that relentlessly gets worse. When Mackenzie was diagnosed, gene therapy for SMA wasn't yet an option. Rachael and Jonny were offered the option of taking part in a trial of treatment with another drug, Nusinersen, but they made the difficult decision that this was too uncertain a prospect, and was too likely to leave their daughter — who was already showing signs of weakness that had led to the diagnosis — living with a poor quality of life. So they launched into making their little girl's short life as full as it could possibly be, giving her all the experiences they could. Snow, sun, sand; every day held something new. Mackenzie's life was joyful — but from the time of the diagnosis, her parents lived with the knowledge that it would be very short, and so it proved.[2]

Mackenzie's doctor was Michelle Farrar, who has been at the forefront of research into the treatment of SMA. When she first met the Casellas, they asked the questions all parents have when faced with news like this. Why did this happen? Then, when they learned that this was a genetic condition, and that they were almost certainly carriers, they asked — why didn't we know? Gently, Michelle explained that almost all parents of children with SMA are in the same boat as the Casellas: they have no family history, and only find out they are carriers once the diagnosis is made in their child. There are screening tests

---

2   Rachael has written a book about Mackenzie's life and what happened next, called *Mackenzie's Mission*. It's a moving personal account and well worth a read.

available that can identify carriers, and which could have told them that this could happen — but you can only have screening if you know it is an option.

The news shocked and distressed Rachael and Jonny, as it must have shocked and distressed many others in recent years, since screening first became available. But right from the beginning, in the midst of their grief, there was something else: determination. This, they decided, was a situation that could not continue — something needed to be done.

So they set out to do it.

Screening for genetic conditions has a longer history than you might imagine, going back more than 60 years. Among the many who have contributed to the story of screening, there are two names of special note — Bob Guthrie and George Stamatoyannopoulos. Of the two, Guthrie is by far the more famous, but perhaps with time the name of Stamatoyannopoulos will become as well known.[3]

I have met perhaps a dozen people with phenylketonuria (PKU). Of those, only one has had the classical features of this rare condition. Frank was a man born in the 1930s, before newborn screening for PKU began. At the time I knew him, he was in his 60s and had never learned to speak. His head was very small, he had a history of seizures, and the staff at the institution where he lived reported that he was often aggressive.

By contrast, all of the other people with PKU that I have met have been well, healthy children or adults, with normal intelligence and no neurological problems. The contrast between them and Frank could not have been more stark, and the reason for it was that they had all been screened for PKU in the first week of their lives. The early diagnosis had made it possible to

---

3   In his later career, Stamatoyannopoulos was a pioneer of gene therapy; his long and illustrious career extended from before the sequence of the first gene was discovered, until a time when gene therapy became a reality.

start treatment immediately, and completely prevent the damage to their brains that otherwise would have been inevitable.

PKU is another metabolic condition, a disorder of the body's chemistry, and it has to do with how the body deals with an amino acid called phenylalanine. Our bodies use amino acids to build proteins, and, in turn, when we eat foods that contain protein, we are consuming amino acids. In most of us, an enzyme converts the phenylalanine we eat into another amino acid, tyrosine. In people with PKU, that enzyme doesn't function properly and levels of phenylalanine soar.

Unfortunately, at high levels, it is toxic to the brain. During pregnancy, the placenta filters the baby's blood and no harm is done, so that babies with PKU are born with perfectly normal brains. As soon as they start drinking milk, they are taking phenylalanine into their bodies and the damage starts.

In the 1950s, a German doctor, Horst Bickel,[4] showed that a diet very low in protein could lower levels of phenylalanine in the blood, with some benefits for people with PKU. Bickel and two colleagues published a paper describing the first successful treatment of a child with PKU.[5] The paper, published in the journal *Lancet* in 1953, makes for somewhat confronting reading by today's standards. Part of that is the bluntness of the language, which to be fair was usual for the time. 'She was an idiot and unable to stand, walk, or talk: she showed no interest in her food or surroundings, and spent her time groaning crying, and banging her head'. The paper describes a series of experiments to modify the food the girl was given, ending up with a diet containing almost no natural protein, tiny supplements of phenylalanine to supply the body's essential needs, and the addition of a special formula that contained all of the other amino

---

4 By a nice coincidence, Bickel and Guthrie shared a birthday, 28 June, which is now international PKU day.

5 Unusually, we know the girl's name: Sheila. She was diagnosed aged 17 months and, reportedly, her mother begged Bickel to try to find a treatment for her. Against enormous odds, Bickel and his colleagues succeeded.

acids, except phenylalanine. The effects were dramatic: the child learned to crawl, then to stand; 'her eyes became brighter, her hair grew darker; and she no longer banged her head or cried continuously'.

So far, so good. But how to be sure it was the treatment that was doing this? Even in 1953, it's clear that Bickel and his colleagues were aware of the risks of introducing a treatment that only *seemed* to work. To make sure, they decided to add a large amount of phenylalanine back into the formula. This is scientifically sound: try a treatment, see what happens, withdraw it, see what happens, start it up again. The part that is shocking to a modern reader is that they *deliberately concealed this plan from the child's mother*. They wanted her to be an unbiased observer of what happened after the phenylalanine was re-introduced. What happened was instant regression: within a day, the girl had lost nearly all the gains of the preceding ten months. Remarkably, the child's mother then agreed (perhaps in reality she was given no choice) to a repeat of the experiment. Again the girl was given the phenylalanine-free formula and learned new skills, again (this time during a hospital admission) the toxic (to her) amino acid was added in, again the skills were lost. The treatment worked. Although Sheila undoubtedly had some permanent brain damage, in the long run, she must have been far better off with the treatment than she would have been without it.

Within a short time, this new way of treating PKU was being widely adopted. Treated children made significant gains, and the younger they were when the treatment was started, the better the outcome. This was because there was always some damage that could not be reversed, and the older the child, the further the damage had progressed. A grim clock was ticking from the moment of birth, counting out the time before it was too late to start treatment. Decades later, in the 1990s, I met Frank soon after his diagnosis was made, and we started treatment. There was some improvement in his behaviour, but that was all: the chance for meaningful benefit had long passed.

Back in the 1950s, the best results of all were in babies diagnosed soon after birth not because of symptoms in themselves but because they had an affected older sibling. From those early trials, the early indication — since proved correct — was that the fate of an affected baby could be utterly changed, with a devastating neurological injury completely avoided.

PKU had become a treatable disease,[6] and the stage was set for Bob Guthrie's rise to greatness.

Guthrie seems to have been something of a difficult character. The tributes written by colleagues after his death are carefully worded, in a way that tells its own story. I asked Bridget Wilcken about Guthrie. Bridget herself is something of a legend in the world of screening — she steered newborn screening in New South Wales through its first 50 years, and has been a world leader in the field for decades. She spoke of Guthrie's great contributions to medicine, but she also told me that, many years ago, Guthrie was a guest at her house. He omitted to tell her, until dinner was already served on the first night, that he was a vegetarian. It just hadn't occurred to him to mention it. This was at a time when it was uncommon to be vegetarian, and as a host you wouldn't think to ask. Others speak of Guthrie calling collaborators at all hours of the night to discuss some new idea, or the details of an ongoing project. Later in the story of screening, he apparently resisted the addition of some other conditions to newborn screening, perhaps because they didn't form part of his vision for the test.

Perhaps, though, that one-eyed focus on the task at hand

---

6   The treatment is very effective, but can be a real challenge for families to manage. The diet is extremely strict, the essential supplements are becoming more palatable but still have a very distinctive taste, and regular blood tests are needed. Especially strict management is needed when a woman with PKU has a pregnancy, because even moderately raised levels can be very harmful to the developing baby. As a side note, if you drink diet soft drinks, have a look at your next can: it may well say 'Phenylketonurics — contains phenylalanine' on the side. The artificial sweetener aspartame contains phenylalanine; even the small amount in a can of Diet Coke could be a problem for someone with PKU.

was the secret of his success. Guthrie was a father of six; his second child, John, had intellectual disability and as a result, Guthrie and his wife became very active in the Buffalo Chapter of the New York State Association for Retarded Children. Through meetings of this group, Guthrie became aware of the existence of PKU and its treatment. One of the difficulties of managing PKU was measuring the levels of phenylalanine in the blood. Guthrie had been working in cancer research for over a decade and realised that it would be possible to adapt a simple test he had been using in his work to do this.[7] He moved to the Buffalo children's hospital and began developing the test. One of his greatest achievements, and one of which he was most proud, was the development of filter-paper cards onto which blood from the baby's heel is dropped. Once the blood has dried, the cards can easily be mailed back to the laboratory that does the testing. If you've had a baby, it's almost certain that he or she was screened, using tests that are only possible because of the existence of these cards, which are still known as Guthrie cards.

Once the test had been developed, it became possible to diagnose babies with PKU in the first weeks of life, before irreversible damage had been done to their brains, and get treatment started. Guthrie made it his mission to promote screening and to make it possible for others to get started. The first trials of screening using this method[8] began in 1960, and, by 1963, 400,000 babies from 29 US states had been screened, with 39

---

7   The test is simple but elegant. Bacteria are grown on agar jelly that contains a substance that stops the bacteria from using phenylalanine. This inability to use the amino acid prevents them from growing — effectively they are being starved. A circle of filter paper soaked with the baby's blood is placed on the agar. If there is a lot of phenylalanine in the blood, it overcomes the effect of the blocking substance, and the bacteria can grow. The more phenylalanine, the better the growth. You put spots of blood from lots of children onto a tray of the jelly, keeping track of which spot came from which baby, of course; incubate; then see which spots (if any) have made the bacteria grow.

8   A less effective urine-based test had already been introduced in some areas.

babies with PKU identified — and saved.

One reason for the rapid early progress of screening was that US president John F. Kennedy had a sister with intellectual disability, and had ensured that the Children's Bureau, which had oversight of this kind of program, was well funded. There are other examples of politicians with a connection to an issue making a difference of this kind; particularly relevant for our purposes is former British prime minister David Cameron, whose son had a rare, severe form of epilepsy, Ohtahara syndrome. Cameron's interest in genetic conditions had a great deal to do with the subsequent developments in genomic medicine in the United Kingdom, which has a world-leading diagnostic and research program.

Over more than half a century, newborn screening has improved and prospered. Most services in developed countries screen for 40 or more different conditions. Not all have such clear-cut benefits from early diagnosis as PKU, but there is no doubt that many tens of thousands of children have had their lives saved, or profoundly improved, by that simple heel prick. Because it seems like such a simple thing, it's easy to take newborn screening for granted. Unless someone in your family is affected by one of the conditions, it's something that barely touches your life. A few minutes of time lost in the flood of events and emotions soon after the birth of a new baby — many people quickly forget the test was even done. But make no mistake, this is one of the great triumphs of medicine.

Newborn screening, wonderful though it is, has the disadvantage that it can only ever be done on a child who has already been born. For many conditions, however, there are no effective treatments yet, and many parents would prefer to have a choice about whether they will have a child with a significant genetic condition. For this reason, screening for genetic conditions during and even before pregnancy has been developed.

Screening for genetic conditions during pregnancy also goes back a long way. As early as 1955, a method had been developed for collecting a sample of amniotic fluid during pregnancy (amniocentesis), and this was initially used for determining the sex of the fetus by examining the cells found in the amniotic fluid. By 1966, the technique had been improved to the point that cells from the amniotic fluid could be grown in the laboratory and their chromosomes analysed. In 1968, a 29-year-old woman who was known to have an inherited rearrangement of her chromosomes[9] that gave her a high chance of having children with Down syndrome attended the Downstate Hospital in Brooklyn, New York. She was 16 weeks into her third pregnancy, having already had a healthy daughter and a son with Down syndrome. An amniocentesis was done and showed that this fetus was also affected, and the woman chose to have a termination of pregnancy. At this time, amniocentesis had already been used to diagnose several different conditions in the fetus with biochemical tests. Now, for the first time, a genetic test had been used to make a diagnosis during pregnancy. The next few years saw rapid development in the field of prenatal diagnosis. Once the technical skills to perform the procedure and the laboratory capacity to do the analysis became widely available, screening of large numbers of pregnant women also became a possibility.

Over the decades that followed, screening for Down syndrome (and, later, other chromosomal conditions) has been performed using ever more sophisticated methods. The goal is to identify pregnancies with a higher chance of being affected, so that an invasive test such as amniocentesis[10] need only be offered to a subset of women to definitively answer the question as to whether the baby is affected. The first screening test was simply to ask the woman her date of birth. The chance of having a baby with Down syndrome, and some other chromosomal conditions, rises with the mother's age.[11] Women are born with

---

9   A Robertsonian translocation, to be exact.

10   Chorionic villus sampling (CVS) is the other main method in use.

11   Men aren't completely off the hook here, because some other types of genetic condition become more likely as fathers age.

all the eggs they will ever have, held in a kind of suspended animation just short of maturity. If you have just turned 35, your eggs are a little older than that ... and their capacity to complete that last step in their development without errors has been declining slowly for years. There's no hard age cut-off at which the chance of having an affected baby suddenly jumps up; the probability rises slowly and smoothly, although the curve does steepen in the mid-30s. For a 20-year-old mother, the chance of having a baby with Down syndrome is about 1 in 1,500; for a 35-year-old, it's 1 in 340; and for a 45-year-old, it's 1 in 32. When I started in genetics, one of the main reasons for having a prenatal test for chromosomal conditions was still 'AMA' — advanced maternal age.[12]

This approach reduces the number of women who might receive more-invasive screening, but the problem remains that, if there is a 1 in 200 chance that an older woman is carrying a baby with a chromosome condition of some kind, the follow-up invasive testing will return 199 negative results for every one affected pregnancy that you identify. When you consider that amniocentesis carries a roughly 1 in 200 chance of causing a miscarriage, you can see that there's a trade-off happening. Are those 199 tests, with their costs and the anxiety they cause, plus one miscarriage, worth identifying one affected pregnancy?

On top of that, screening only women aged 35 and above will miss most affected pregnancies, because, even today, the great majority of babies are born to younger women.

There have been various efforts to address these problems. The first worked by measuring combinations of several substances in the mother's blood that change during pregnancy, with levels that tend to be higher or lower if the baby has a

---

12 I once put my foot in it very badly by wishing one of our genetic counsellors 'happy AMA day' on her 35th birthday. The term 'advanced maternal age' is unfortunate, considering that 35 years is not really an advanced age in most other contexts. There's an even worse term in obstetrics — an 'elderly primigravida' is a woman who becomes pregnant for the first time at 35 or older. 'Elderly'? What were they thinking?

chromosomal problem. The results of these tests, combined with the mother's age, were used to refine the risk assessment. Later, measurement of the thickness of skin at the back of the neck was added into the calculation. Many different medical conditions, including Down syndrome, can lead to a build-up of fluid in this area early in pregnancy, so the measurement improved the accuracy of the screen.

Every version of these tests still suffered from the same basic problem. Even as they became more sensitive, with lower false-positive rates, that trade-off remained; for most women who were flagged as being at higher risk, it was still much more likely that, if they chose to have an amniocentesis, the result would be normal — yet they still were exposed to the risk of miscarriage from the procedure. In that sense, having a screening test carries risks — the risk that you might have an invasive test you don't need, and possibly even lose the pregnancy.

Recently, a far better screen has become available.[13] Non-invasive prenatal screening (NIPS) uses the new genetic sequencing technology in a nifty way: it treats DNA simply as something to be counted. Usually, when we extract DNA from a blood sample, we take it from the white blood cells, each of which has a nucleus (red blood cells have lost theirs, and their mitochondria, so they don't contain DNA). However, there is a small amount of DNA in the plasma, the fluid that makes up the half or so of your blood that isn't blood cells. In the late 1990s, it was discovered that, if you take blood from a pregnant woman, discard the cells, and extract DNA from the plasma, some of that DNA will come from the placenta. It's often called the fetal fraction, but the fact that the DNA is from the placenta, not directly from the fetus, is important. We'll see why shortly.

---

13   It's a bit harder to pin down who the key people were for some of the prenatal screening methods, because of contributions from many different people over a long period of time. But Dennis Lo from Hong Kong was the first to develop NIPS, and Kypros Nicolaides from London led the use of ultrasound for Down syndrome screening.

There are different ways that NIPS can be done, with several companies having come up with their own approaches. The most common approach involves sequencing DNA using one of the new sequencers that reads many individual strands of DNA at once (the same ones discussed in chapter 5). Sequences are read from regions spread across the genome. Each continuous stretch of sequence, representing one molecule of DNA, is a 'read'. The fetal fraction varies between blood samples, but suppose for example that it is 9 per cent. On average, you expect that, at any one place, there will be 100 reads from the mother's DNA and ten from the placenta (representing the fetus); ten out of 110 is 9 per cent. You don't necessarily even have to distinguish which are which: all you need to do is count. If you have, on average, 110 reads for every chromosome except 21, but 115 reads[14] across chromosome 21 — then there are half again as many from the placenta as you expect: there must be three copies of that chromosome instead of two. The baby has Down syndrome.

Except if it doesn't.

There are various ways you can be fooled into thinking the baby has a chromosome abnormality when really it doesn't. One of the most important reasons[15] this can happen is that, sometimes, there are chromosomal changes in the placenta that are not in the fetus. The early embryo is a ball of cells that splits into two: part of it goes on to become the fetus, and eventually the baby, and the rest becomes the placenta, membranes, and so on. If a mistake in cell division happens after that split, the placenta can have a chromosome change it doesn't share with the fetus ... and since the fetal fraction is really a placental fraction,

---

14   You'll likely need to do more sequencing than 110 reads for the statistics to work, but the principle remains the same.

15   Most of the other reasons are technical, but a rare cause is that the mother has cancer, the cancer cells have abnormal chromosomes, and they in turn are spilling DNA into her blood that has abnormal chromosomes. There have been a number of women around the world who have unexpectedly had a cancer diagnosis made this way.

the test can be fooled. It really is correctly counting an extra copy of chromosome 21 — it's just that the extra copy is in a tissue where it doesn't matter.

This means that this truly is a *screen*. It's often called 'non-invasive prenatal testing' (NIPT), but I strongly prefer 'non-invasive prenatal screening' for this reason. There is good reason to believe that some obstetricians have misunderstood NIPS reports and acted on that result alone, without doing a confirmatory test.[16] If so, there would certainly have been some terminations of pregnancy that were done based on a mistake, with the baby not having had any chromosomal abnormality at all — a disturbing thought.

Don't get me wrong — NIPS is *very* good at what it does. It misses far fewer affected pregnancies than the other screening options, and a positive result is much less likely to be wrong. In the time since NIPS first became available, we've seen a sharp fall in the number of prenatal tests being done, thanks to this better performance. Recently, however, the trend has been re-versing, because of mission creep.

Most NIPS is provided by private companies, and there are limited ways that they can compete with each other. Price is one option, obviously, but making the case that their test is better in some way is another. One way to do that is to test for more conditions than the competition. NIPS started out looking for extra copies of chromosomes 13, 18, and 21 as well as extra or missing copies of the sex chromosomes. As we've seen (in chapter 4), testing for sex chromosomes has the potential to produce results that are not straightforward. Still, overall, this approach produced a relatively small number of false-positive results — more for chromosomes 13 and 18 than for 21, but, even so, a NIPS result that says there is a high chance that the fetus has an extra copy of chromosome 13 has a reasonable chance of being correct.

---

16   Ideally the confirmatory test would be an amniocentesis, because chorionic villus sampling also looks at the placenta rather than directly at the baby.

Adding in extra targets has been … problematic. Some companies have added in screening for relatively common, but still rare, abnormalities, such as the deletion on chromosome 22 that causes velocardiofacial syndrome (or Sedláčková syndrome if you prefer). Others have taken to reporting differences affecting other chromosomes if they happen to show up in the data, such as an extra copy of chromosome 10.

One of my jobs in the lab is to issue reports about prenatal diagnostic tests. So far, I have only seen one 'extra target' NIPS result turn out to be a true positive that we confirmed at amniocentesis, a large chunk missing from one chromosome. All of the follow-up tests for smaller deletions that I've reported on so far have been negative, although presumably at some point that will change.

The reason for all these false positives is mainly a simple quirk of statistics. It's one that can affect all sorts of laboratory tests, and it can be summed up like this: the rarer the condition, the more likely a positive test result for that condition will be wrong. Unintuitively, the exact same test can give you different results, depending on whom you are testing.

NIPS test reports often include a statement similar to: 'this test is 99.9 per cent sensitive and 99.9 per cent specific for the detection of Down syndrome'. That sounds very impressive, and it is: 99.9 per cent sensitive means that if 1,000 women who were carrying a baby with Down syndrome had the test, 999 would have a positive result, and only one would be missed. That's *really* good, way better than the best of the previous options, which would detect 900 and miss 100, rather than just one. The problem lies in the other number, the specificity: 99.9 per cent specific means that if 1,000 women carrying a baby *without* Down syndrome had the test, one would have a false positive.

One out of a thousand doesn't seem so bad, does it? To see why this isn't as good as it sounds, here are some simple numbers that are made-up but make the point. Let's say you have two groups of women. Group A have a 1 in 100 chance of having a baby with Down syndrome, because of their age. Group B are younger and only have a 1 in 1,000 chance. A thousand women from both groups have NIPS done, using a test like the one described above.

In the 1,000 women from Group A, with a 1 in 100 chance, there are 10 affected pregnancies. The test is so sensitive that all of them are detected. There is also one false positive (the 1 in 1,000 chance comes up). That means there are 11 positive results, of which 10 are correct. For these women, a positive result has a 10 out of 11 chance (91 per cent) of being correct.[17]

Now we test 1,000 women from Group B. There is one affected pregnancy, with a (correct) positive result. There is also one false positive. For these women, a positive result has only a 1 in 2 chance (50 per cent) of being correct. A positive result from exactly the same test has a different meaning in the two groups.

As you can see, the rarer a condition is in the population that's being tested, the lower the chance that a positive result will be correct. If the specificity is a bit lower (because you are looking for a smaller target than a whole chromosome, for instance) things will get worse. Because the tests were originally designed to look for the most common problems, each extra condition you add is going to be rarer than the last, and the test will perform worse as a result. In the last couple of years, we've started to see the number of invasive tests in pregnancy rise again. Is more better? Is it a good idea to extend these tests out to look for rarer conditions? If you get a true positive for a rare condition, giving you options you wouldn't otherwise have had, you might think it is. If you have a miscarriage due to an invasive test that was done because of a result that then turned out to be a false positive, you may think otherwise.

---

17   This is called the positive predictive value (PPV) of the test.

At about the time Rachael and Jonny Casella were learning the terrible news about their daughter Mackenzie's diagnosis, the parents of two other children who were patients of Sydney Children's Hospital were hearing similar news, and they were hearing it from me. Twice in the space of ten days, I had much the same conversation, gave the same stark message. We know now why your child has been having these symptoms. This condition is not curable, it will steadily get worse, and within a few years it will be fatal.

No matter how carefully or kindly, no matter how well you deliver news like this, you know that it is a hammer blow. For the parents, this will forever be remembered as one of the worst days of their lives.

At some point — often in the very hour that they learn the diagnosis — parents who receive news like this will ask the same question that Rachael and Jonny asked. Why did this happen to our child, to us? Wasn't there something that could have been done? These days, they have often had screening for chromosomal disorders during the pregnancy, and it's common that they mistakenly think that this was a test for *all* genetic conditions. So there's another question — we had all the tests, why wasn't this picked up?

For almost everyone who has a child affected by an autosomal recessive condition, there is no family history, no prior warning of the risk. For X-linked conditions, there may be a family history, but often there isn't. This means that the only way to find out if you're a carrier before having an affected child would be to have a carrier screening test. For most of my career, there simply haven't been such tests available for most conditions, for most people. When I gave parents this type of bad news, I could at least look them in the eye and tell them there was no way they could have known in advance that this could happen to their child.

Now, though … things are changing.

Several years ago, when it first became evident that massively parallel sequencing was going to become available for reasonable prices, I started to think about carrier screening in earnest. In that, I was well behind the times.

Usually, being a carrier for a recessive condition is neither here nor there. It does you no harm, but it also does you no good. There are exceptions, however, and the best known of these has to do with malaria and a group of conditions that affect red blood cells. The various forms of the malaria parasite (species from the genus *Plasmodium*) have a life cycle that shuttles back and forth between humans and mosquitoes. In humans, the parasite spends part of its life living inside red blood cells (which are consumed by mosquitoes, which subsequently infect other humans, and so on). There is a group of blood conditions, the thalassaemias, in which the oxygen-carrying protein, haemoglobin, is abnormal, and in turn the red blood cells that contain haemoglobin are abnormal. They are fragile and don't last very long in the bloodstream after they form; at worst, affected children would be lucky to make it to their third birthdays unless they have regular blood transfusions.

For carriers, though, the news is rather better. Their mildly fragile red cells do a good-enough job of carrying oxygen through the body, and usually this causes no health problems at all. But from the point of view of a *Plasmodium*,[18] those mildly abnormal cells are an uncomfortable place to be. This gives partial protection against the effects of malaria, and especially against its most severe forms. Malaria remains a serious killer: according to the ongoing Global Burden of Disease study, 619,827 people[19] died of malaria in 2017, out

---

18   To the extent that a *Plasmodium* has a point of view.

19   Yes, this is a suspiciously precise number. In case you're interested, the top three killers were cardiovascular diseases with 17.8 million, cancers with 9.6 million, and respiratory diseases with 3.9 million. Malaria killed more people than many other causes, including murder (405,346), drowning (295,210), terrorism (26,445), and disasters (9,603). Malaria killed 23 people, mostly small children, for every one who died due to terrorism in that year. This is why the front page of your newspaper is forever running banner headlines about the ongoing malaria disaster. It isn't? ... nor mine, for some reason.

of 56 million deaths in total. Reduce your risk of death from malaria, and you increase your chances of living long enough to have children, and pass on your genes to the next generation. That's called a selective pressure — if people with a particular type of genetic variation are more likely to successfully reproduce, that variation will become increasingly common in the population.

The mosquitoes that can carry malaria like it hot, or at least warm: before people started trying to do something about it, malaria was found as far south as 32° of latitude, and as far north as 64° (!) ... but it was always most concentrated in a belt around the equator, and has largely remained so. Where malaria is or was, you can expect to find thalassaemia and related conditions, and carrier frequencies can be very high indeed.[20] This includes most of the countries around the Mediterranean — which brings us to Italy, to Greece, and then to Cyprus.

In 1955, the Italians Ida Bianco and Enzo Silvestroni suggested the possibility of preventive counselling: if you identify couples who are carriers, you could counsel them against having children. George Stamatoyannopoulos, fresh out of medical school, took on the challenge. In 1966, he went to Orchomenos, a village in Greece with a population of 5,000, and started screening for sickle cell disease (a variant of thalassaemia that is common in Africa but also crops up in some other places, including parts of Greece). There were a lot of carriers: nearly a quarter of the population. About 1 in 100 babies born in the village was affected. Stamatoyannopoulos advised unmarried carriers to steer clear of each other and choose non-carriers to marry, but, when he returned to the

---

20 You might think that a condition that is fatal when you have two copies of a faulty gene would be prevented by that from becoming common in the population. Consider though, that if 1 in 10 people is a carrier for a condition, in only 1 in 100 couples are both partners carriers (1/10 times 1/10), and therefore only 1 in 400 children are affected (1 in 100 times the 1 in 4 chance for each baby born to a carrier couple of being affected). If there is a benefit to carriers, 1 in 10 people benefits, but only 1 in 400 suffers negative consequences.

village, he found that his advice had been ignored. In that sense, the effort had not been a success — but the first attempt at carrier screening for reproductive purposes had been made, and a great deal had been learned. In 1971, when the World Health Organization called on Stamatoyannopoulos to visit Cyprus and advise about the problem of thalassaemia on that island, he was well prepared.

The situation in Cyprus was not unlike that of Orchomenos, although it was at the scale of an entire country rather than a single village. The carrier frequency for thalassaemia and the frequency of affected babies were just a little lower than for sickle cell disease in Orchomenos, but the impact on peoples' lives and on the health system was considerable. Nearly half of the output of the blood bank in the capital, Nicosia, was being used for keeping people with thalassaemia alive, and, at the end of that decade, 6 per cent of the entire budget of the Ministry of Health was being spent on a single drug, desferrioxamine, which is needed to treat people for the dangerous overload of iron that comes with frequent blood transfusions.

It took much of the 1970s to figure out how best to do carrier screening and get it working effectively. An early effort to persuade carrier couples not to marry each other was just as unsuccessful in Cyprus as it had been in Greece. By 1977, it had become possible to do prenatal diagnosis, and attention was focused on screening people of reproductive age in order to give them information on which they could consider acting. This won strong support from the Church of Cyprus, because of the realisation that there were *fewer* terminations of pregnancy as a result of prenatal testing. Up to that point, many people who knew they had a 1 in 4 chance of having an affected baby were choosing termination of pregnancy rather than take that chance. For those couples, prenatal diagnosis meant that three out of four pregnancies could be shown to be unaffected, and thus would continue.

In 1979, the expected number of affected babies born in Cyprus (based on historical figures) was 77, and the actual number was only 18. Given a choice, couples were choosing to take steps not to have affected children.

The story of carrier screening in the decades since then has been one of mixed successes and mostly slow progress, until a recent boom. Carrier screening for the thalassaemias and related conditions, targeted at people from populations where these are common, is cheap and generally quite effective in many countries around the world. There are targeted screens for some other conditions, too.

In this regard, Israel is the undisputed world leader. There are a number of recessive conditions that are more common in people of Jewish ancestry, especially the Ashkenazi Jews (those who trace their ancestry to Central Europe). Most infamous of these is Tay-Sachs disease, another lysosomal storage disease that affects the brain. In the classical form of Tay-Sachs, most affected children do not reach their fourth birthday. There are community-led screening programs for Jewish people in various parts of the world, but in Israel the Ministry of Health offers free screening to everyone who is planning a pregnancy or is early in a pregnancy. It's all targeted, to a very fine level of detail. There's a list of genes recommended for most of the population, then different sets depending on ancestry — for Ashkenazi Jews, for Jews of North African origin (except Morocco), for Jews of Moroccan origin, and so on; and also for those from other populations: there's a specific list just for Bedouins in the Negev region, for instance.

In much of the rest of the world, in order to access carrier screening, you need two things that not everyone has: information and money. You need to know that the tests exist, and you need to be able to afford to pay for them (or have health insurance that will pay for them).

Currently available tests range from covering as few as three conditions to as many as hundreds. 'Three conditions' in this case usually means SMA, which we've encountered already, cystic fibrosis (CF), and fragile X syndrome. CF is a complex condition in which the body's secretions are thicker than they should be. That might not sound so bad, but it really is — untreated, this causes children to have progressive, serious lung infections and severe nutritional deficiencies due to failure of the pancreas; in the past, most affected children did not make it to adulthood. Modern treatment makes a lot of difference to the condition, with greatly improved life expectancy, but is burdensome on child and family. Fragile X syndrome is a common cause of intellectual disability. If someone is offering a screen for hundreds of conditions, they will almost always include these three.

This was the news that shocked Rachael and Jonny Casella. Had they known about SMA carrier testing, they could easily have found out that they were carriers before having a pregnancy — opening up choices, including PGT and prenatal diagnosis.

They became powerful advocates for screening, starting by writing to every member of the Australian Federal Parliament, as well as to New South Wales state politicians. They met with the state health minister, with the federal deputy health minister, and finally with the federal minister for health, Greg Hunt.

By a remarkable and fortunate coincidence, at much the same time that the Casellas began their advocacy, a group of researchers were talking to the government about the same topic. Nigel Laing, an internationally renowned expert on the genetics of muscle diseases who had advocated for carrier screening for many years, had convened a meeting of interested Australian doctors and scientists in late 2016. I attended the meeting and spoke about some research I had led, studying screening in couples who were related to one another. Because we share genes with our relatives, such couples have a much higher chance of having children affected by recessive conditions than those who are unrelated. We showed that screening using a very large panel of genes, about a quarter of the exome, worked well in such couples and was acceptable to them.

Thanks to Nigel's leadership, the nucleus of an Australian team of carrier screening researchers was already in place in 2017, when I had to give terrible news to parents twice in ten days. That experience prompted me to reach out to the group, and together we decided to write to the federal health department to suggest that this was an area which needed attention.

A series of meetings in Canberra followed. There was support in principle within the health department for a pilot project, but nothing concrete was on the table. Unknown to us, however, we had a secret weapon: the Casellas. Michelle Farrar, Mackenzie's neurologist, introduced me to Rachael and Jonny, and they arranged for us to be in the room when they met with Minister Hunt. The Casellas spoke about Mackenzie's life and about their loss. They urged action on carrier screening. It was obvious that the minister was deeply moved by their story, as was everyone in the room. Hunt, already a strong supporter of research into genomic medicine, promised that he would take action, and he has been as good as his word. The government committed $20 million to a research project aimed at determining how best to introduce carrier screening to Australia, with the goal that screening should be available free of charge to any couples who wish to access it. The project was christened Mackenzie's Mission by Hunt.

I am co-leading the study with Nigel Laing and Martin Delatycki,[21] an eminent geneticist from Victoria who is a long-time advocate for screening for genetic conditions. We plan to screen 10,000 couples over the course of the study, with research into every aspect of carrier screening. You might think there are few questions left to answer about such a simple concept, but it's surprising how many details remain to be worked out about how best to deliver screening.

A major task in the first year of the project was to work out what, exactly, we should be screening for. This might seem easy enough — it's a matter of choosing severe genetic conditions and screening for those, right? But straightaway that introduces difficulties. What do you mean by 'severe'? There are plenty of

---

21   Working with Martin and Nigel, and with the many, many others involved in the project, has been a wonderful experience. It seems like nobody ever says no to helping with Mackenzie's Mission.

conditions for which this is easy. Uncombable hair is an autosomal recessive condition;[22] few people would consider that severe enough to screen for. The information just wouldn't be useful to most people — certainly it's unlikely that many people would change their reproductive decisions on that basis. At the other end of the spectrum, a lethal condition like Tay-Sachs disease is also straightforward; almost everyone who thinks screening is a good idea would include that. But there are many conditions that sit in a grey zone, where some might think them sufficiently severe but others do not. Take deafness, for instance. Most if not all of the commercial screening tests include at least one form of deafness. But how severe a condition is it, really? Some think that it should not be considered a 'condition' at all — it has been argued that treating deafness is a form of cultural genocide, because it may eliminate the Deaf community and its languages.

For Mackenzie's Mission, we decided that we would include genes if the associated condition caused a medical problem that started in childhood, that was severe, disabling and/or life-shortening without effective treatment (or where treatment is very burdensome), and for which an average Australian couple would take steps to avoid having a child affected by that condition. After much debate, we concluded that deafness does not meet those criteria, and we have not included any genes for isolated deafness.[23] However, we think the questions of what to include in a screening program and where the boundaries should be drawn deserve further work, and one of the questions we aim to answer during the course of the project is whether Australians as a group agree that we have this right.

Our final gene list wound up being much longer than we had expected — 1,300 genes associated with more than 700

---

22   While not affected by this condition, I am nonetheless undergoing the cure — like many men my age.

23   In other words, deafness where that is the entirety of the condition. Deafness can be part of syndromes, too, in combination with a variety of other medical problems.

conditions. There are more genes than conditions because there are plenty of conditions that can be caused by variants in multiple different genes. Of course, many of these are very rare, and for the majority there will be no couples among the 10,000 we screen who are found to have a 1 in 4 chance of having an affected child.

Other research questions we need to answer include some quite simple ones — for instance, how many of the 10,000 couples will be identified as carriers for one of the conditions? That one is simple, but important — if you want to plan a population-wide screening program, you need to know what resources will be needed, and this is a key piece of information. A similarly pragmatic question is: will screening be cost-effective? This may seem callous given the human impact of genetic conditions, but, if a government is going to pay for screening, it needs to know if it can afford it, so health economists are a key part of our team. Other questions are more complex: What are the ethical implications of doing this type of screening? How can we design a program that does the most good and the least harm? How do we translate the research evidence we generate into medical practice as efficiently and effectively as possible?[24] And so on.

There's no such thing as a perfect screening program. There will always be gaps in our knowledge and limits to our ability to identify those who are affected by, or carriers for, genetic conditions. I'm hopeful, though. I hope that within a few years, we will be able to offer the option to be screened to all who wish it. I hope that many will choose to be screened, and that, while most will receive reassuring information, for those who are found to have a high chance of having an affected child, the information will be helpful.

Most of all, I hope that, over time, I will need to have fewer meetings with young couples to give them bad news about their children.

---

24 This is a whole field of academic effort, called implementation science.

There is a great deal to be optimistic and excited about in genetics. This book has been about the past and present of genetics, and how it affects peoples' lives. But what about the future?

Where to from here?

To complain of the age we live in, to murmur at the
present possessors of power, to lament the past, to
conceive extravagant hopes of the future, are the
common dispositions of the greatest part of mankind.

EDMUND BURKE (1770)

In a room right next to my office in the lab, there sits an extraordinary machine. It is the NovaSeq 6000, which is capable
of sequencing the genomes of 6,000 human beings per year. Its
manufacturers, Illumina, used to make sequencers that looked
like they would fit just fine as props in a science fiction movie.
With the NovaSeq, however, they seem to have given up on
industrial design, and the device looks uncannily like a large
washing machine. The message is clear enough — when you
have this much power, you don't need to *look* impressive.

Six thousand genomes per year. One machine, and it is one
machine among many — there are three others in New South
Wales alone, and likely hundreds or even thousands worldwide.
When I started my career, we hadn't even sequenced a single
human genome. Back then, the idea that I might be sitting at a
desk just a couple of metres away from a device that could do
so much would have seemed wildly implausible.

This suggests that any forecasts I make about the next 20

years, or even the next five, should be viewed as somewhat suspect. But there are some pretty safe bets.

The first safe bet relates to technology. It will keep getting better, cheaper, and faster. Ten years from now, the NovaSeq 6000 will be a sadly outdated piece of equipment, and few labs will still be using them. At some point in the coming years, the costs for whole genome sequencing will come down so much that there won't be any point in doing exome sequencing any more. Already, as we saw in chapter 10, we sometimes sequence all 23,000 genes in order to get information about just one or two, and it's quite likely that genome sequencing will replace many tests we currently do that are targeted at particular changes. It's not impossible that, one day, if we want to find out information about a single base of someone's DNA, we will read all three billion and just look at that one position we are interested in, ignoring everything else.

The next safe bet is that genetic testing will become more and more commonplace. Exome sequencing has already gone from being an exotic, very expensive test ordered only by clinical geneticists to being a routine test ordered by many different specialists. Within a couple of years, attitudes among my colleagues have gone from 'wow, my patient is going to have her exome sequenced' to 'why is this exome result taking so long to come back?' We will be doing more and more complex testing, and it will be done earlier. There's a term used to describe the often years-long process by which people used to reach a diagnosis for a rare genetic condition (if a diagnosis was ever made): the diagnostic odyssey. Many different tests, many appointments with specialists, years of frustration and uncertainty … often, now, we can avoid that altogether by doing exome sequencing as soon as it is apparent that there is a severe problem. The diagnostic odyssey should quickly become a thing of the past.

Along with improving access to better, faster testing, the situations in which it is used will broaden. Already, there are large research projects involving sequencing the genomes of cancers to look for genetic changes that might respond to specific treatments. The term 'precision medicine' is used to describe the use of genetic testing — be it for cancer or other conditions — in order to tailor a treatment to a patient's individual genetic make-up. The term itself is something of a meaningless buzzword, because it ignores so much that has gone before. I would argue that identifying a child with PKU using newborn screening, then giving that child a precisely targeted treatment that lets her avoid the devastating effects of the condition, deserves the name 'precision medicine', if anything does. If you have a bacterial infection and the lab tests the bacterium that ails you to work out exactly which antibiotic it is sensitive to … that seems precise, too. So I prefer the term 'medicine' to describe the new advances. *Medicine* is going to get better, and more genetic. Perhaps we will even get to the point where general practitioners will order whole genome sequencing before referring patients to a specialist.

But that leads to a problem that doesn't have an obvious end in sight: the problem of interpretation. Already, the hardest part about exome sequencing isn't generating the data — it's understanding what it means. No doubt we will improve with time, but, just because your doctor may be able to send your blood to have your genome sequenced, it doesn't mean she will necessarily be able to understand the results in a way that will help with your medical care. As we've seen, most human disease is genetic in some way, but mostly this means complex disease, with genetics that are very difficult to interpret meaningfully for an individual. Even when we're looking at variants in a single gene that are known to be linked to a specific genetic condition, it can be a challenge to be sure if the changes you find are the cause of the problem or not. Be wary of the company that offers to sequence your whole genome and interpret the results in a way that will guide you in making decisions about your health: they may be promising more than they can deliver.

Having said that, treatments for genetic conditions are going to become more and more common, and more and more effective. Some will be cures. Most will continue to be eye-wateringly expensive.

Regardless of the changes in technology; regardless of the treatments that may or may not be around the corner; no matter how fast we accumulate knowledge, and turn uncertainty to confidence — the fundamental nature of my field will not change. It always has been, and always will be, *human* genetics. The story of the future of human genetics will be like the story of its past, and its present — it will be the story of people, people like those you've read about in these pages. The scientist, on fire with a new insight. The doctor, working patiently to gather knowledge about a rare condition. The child with a syndrome, growing and living in a way that is defined — even more so than the rest of us — by her genes. The parents of that child: loving, grieving, learning, hoping. It is always, and only, about people.

Here's my final and favourite prediction: there will be things that happen in genetics in the next few years that will be a complete surprise. I have no idea what they will look like, but I can't wait to find out what's next.

Glossary

Amniocentesis: A type of prenatal diagnostic test. Performed under ultrasound guidance, typically at around 15–16 weeks, although it can be a little earlier or a lot later in pregnancy. A long needle is passed through the woman's abdomen, and a sample of amniotic fluid is taken. Amniotic fluid is the liquid in which the baby floats. It contains cells that come from the baby. Nowadays, most testing on amniotic fluid is **DNA**-based and involves directly extracting DNA from these cells or growing (culturing) them in the lab, then extracting the DNA. They can then be used to test for chromosomal abnormalities or other genetic conditions. Old-school **chromosome** analysis can also be done on the cells (looking at the chromosomes under the microscope), as can other types of testing. For instance, biochemical testing can be done on the cells, or on the fluid itself. These are becoming much less common, however.

Autosomal dominant: A form of inheritance in which there is a change in a **gene** on one of the **autosomes**, i.e. chromosomes 1–22, and this is sufficient to cause a genetic condition. You can think of this as the faulty copy of the gene 'dominating' over the other, 'normal' copy. A person who has an autosomal dominant condition has a 1 in 2 chance of passing this on to each of his or her children. Dominant conditions generally affect males and females equally, although there are some that mainly affect one or the other sex. For example, familial breast and ovarian cancer due to **variants** in the genes *BRCA1* or *BRCA2* is a condition that does increase the risk of some cancers (including breast cancer) in men, but mainly affects women. Dominant conditions tend to vary a great deal between affected people, even people in the same family who share the same variant in the relevant gene.

Autosomal recessive: A form of inheritance in which there is a change in both copies of a **gene** on one of the **autosomes,** i.e. chromosomes 1–22. If a person is affected by an autosomal recessive condition, it is nearly certain that both of his or her parents is a carrier for the condition — i.e. they have one faulty and one 'normal' copy of the gene. There are some very rare circumstances in which one (or, in theory, even both) parents is not a carrier but a child is still affected. For example, there could be a new change in the copy of the gene passed on by the parent who is not a carrier. This is something that happens very rarely for most genes, but there are exceptions; **spinal muscular atrophy** is one condition in which — although it's far from common — we do sometimes see affected children who have only one parent who is identified as a carrier. Virtually everyone is a carrier for one to several autosomal recessive conditions. This is almost never an issue for the carrier's health, although again there are rare exceptions.

Autosome: Any of the **chromosomes** other than the X and Y. Chromosomes 1–22 are all autosomes.

Centromere: Part of a **chromosome**. Visible in pictures of chromosomes as the 'waist' of some chromosomes, although the centromere can be at one end of the chromosome as well (this is called an acrocentric chromosome, because the centromere is at the peak ('acro') of the chromosome. The centromere has an important role in cell division.

Channel: The cell membrane is carpeted with **proteins,** or complexes of multiple proteins, which function as channels. These allow specific substances, such as potassium ions, to pass across the cell membrane. Sometimes this is a passive process and sometimes an active pumping. Normal functioning of these channels is important for maintaining the right mixture of salts inside and outside cells and for regulating the electrical activity on the surface of cells.

Chorionic villus sampling (CVS): A type of prenatal diagnostic test in which a sample of the placenta is collected — specifically, the chorionic villi. Typically done at around 11–12 weeks of pregnancy. Under ultrasound guidance, a needle is passed through the woman's abdomen, or sometimes a flexible tube is passed through the placenta, and suction is used to collect the sample. The idea behind the test is that, early in embryonic development, there is a split between the cells that go on to form the fetus, and ultimately the baby, and those that form the placenta, but they start from the same point and share their initial genetic make-up. Genetic test results from the placenta are usually an accurate representation of the fetus, but sometimes there can be changes (especially chromosomal abnormalities) that occur after the split. This usually shows up as apparent chromosomal **mosaicism** at CVS. If there is mosaicism that is only in the placenta, it's called confined placental mosaicism. Usually, this is harmless, although occasionally it can affect the function of the placenta. This means that, if we find a mosaic chromosomal abnormality at CVS, we often have to follow up with an **amniocentesis** to work out the significance of the result.

Chromosome: Chromosomes are structures found in the nucleus of the cell. They are made of very long strands of **DNA**, wrapped around **proteins** called histones. The human genome is arranged into 23 pairs of chromosomes (in most people); you inherit one of each pair from each of your parents. The chromosomes are numbered from 1–22 (the **autosomes**) plus the X and Y chromosomes (the **sex chromosomes**). If you have missing or extra chromosome material, that means you have missing or extra copies of **genes**, which can cause chromosomal disorders.

Coding: DNA that is coding is translated into proteins; non-coding DNA is not. Within a **gene**, the coding parts are the **exons**, whereas the non-coding parts are the **introns** as well as regulatory **sequences** before and after the exons (the 5', pronounced 'five prime', and 3' untranslated regions). Non-coding DNA may still be transcribed, i.e. copied to **RNA**. The resulting RNA molecules may have a variety of functions, including signalling and regulating the action of genes.

de novo: Occurring in a child but not in his or her parents. A new change in the DNA.

DNA: Deoxyribonucleic acid — the stuff of life. DNA consists of a long chain of individual bases — adenine, cytosine, guanine, and thymine. These have a sugar backbone, deoxyribose; DNA forms a chain by linking deoxyribose to deoxyribose, with a phosphate group between each. Two strands of DNA form a double strand, with links between the bases; C joins to G (with a triple bond), and A joins to T (with a weaker, double bond).

Dominant: see **autosomal dominant.**

Enzyme: An enzyme is a type of **protein** that functions as a catalyst — it makes a chemical reaction happen much faster than it otherwise would. Our lives depend on the continuous action of numerous different enzymes.

Exon: the parts of a **gene** that are translated into **protein.**

Gene: A gene can be thought of as a set of instructions to the cell, telling it how to make a **protein.** Genes are long sections of DNA that have a specific structure — regulatory **sequences** (some of which can be a long way away from the gene itself, some of which are immediately upstream and downstream of the coding part of the gene), **exons**, and **introns.** Exons are parts of the gene that are translated into protein. Introns sit in between the exons and are not translated, but they may be transcribed — copied to **RNA**, which then can have a function including regulating the action of the gene. There are some single-exon genes, which do not have any introns. There are also RNA genes that are transcribed to RNA but not translated to protein.

Genome: All of an organism's genetic material. Every living organism has a genome.

Genome-wide association study (GWAS): A type of genetic experiment aimed at finding genetic variation that influences a human characteristic. Large numbers of people who are known to be affected by a condition or about whom you know something (such as their blood pressure) are tested for thousands of variations spread across the **genome**, looking to find a link between a **variant** and the characteristic you are interested in.

Gonadal mosaicism: This is **mosaicism** that occurs in the gonads (ovaries or testicles).

Human Genome Project (HGP): The great project to **sequence** the whole of the human genome.

Intron: The parts of a **gene**, in between the **exons**, that are not translated into **protein**.

Lysosome: A type of **organelle** that is responsible for waste disposal and recycling in the cell. If any of the enzymes in the lysosome do not work properly, the compound it is supposed to recycle builds up inside the lysosome, with harmful effects.

Mitochondria: The mitochondria have many different functions, but perhaps the most important is metabolising digested food (carbohydrates and fats) to produce a form of energy that the cell can use for its various functions.

Mosaic/mosaicism: If there is a genetic change that is present in some cells but not others, this is mosaicism. This can be at the level of a whole **chromosome** or at a much smaller scale, including a single base of **DNA**. In one sense, we are all mosaics, because of the errors that happen whenever a cell divides. For this to be medically important, the mosaicism has to affect a substantial proportion of cells in a given tissue.

Mutation: see **variant**.

Non-coding: DNA that is not translated into **protein** is non-coding. See **coding** for more detail.

Pre-implantation genetic testing (PGT): Also known as pre-implantation genetic diagnosis (PGD). In-vitro fertilisation is used to make embryos that are then tested for genetic conditions. These could be single-**gene** conditions or chromosomal abnormalities. The idea is to implant only an embryo that is not affected by the condition being tested for.

Protein: Proteins are like verbs in the language of the body: if the cell needs something done, it calls on a protein. Proteins can be machines — the strength in your muscles comes from an interaction between a group of proteins that can turn energy into movement. Proteins can be pumps — the **channels** in cell membranes are proteins, or protein complexes. Proteins can be factories — the mechanisms for making new proteins include many proteins; the work of the **mitochondria** in converting food into energy requires the work of many different proteins. **Enzymes** are proteins. Also, though, proteins can be structural components. The collagens that hold your body together are proteins. Proteins are made up of 20 different amino acids (there's also a rare 21st amino acid, selenocysteine — see the Notes for chapter 1 for more on this). To make a protein, **genes** are transcribed (see **transcription**) to a type of **RNA** called messenger RNA (mRNA). The **introns** are cut out (a process called splicing) to make a mature mRNA. This is then translated to protein — structures called ribosomes read the mRNA and add new amino acids to the growing chain. Many proteins then undergo further modification — bits can be carved off each end, chemical changes can be made, and various other substances such as sugars added on — before the protein is fully functional.

Reference sequence: A 'standard' **sequence** of **DNA** for an organism. The human reference has been updated a number of times as gaps are filled in and errors are corrected. The data on which the human reference sequence is based are from numerous individuals, all anonymous, so that it does not represent any one person. It is not the 'right' sequence, but, for the great majority of locations in the genome, it does represent the most common version. If there is a C at a particular place in the reference, it is likely that most people have a C there.

RNA: Ribonucleic acid. Chemically very similar to **DNA**, with the exceptions that the sugar backbone is ribose rather than deoxyribose, and there is a different base, uracil, in place of thymine. RNA has many different functions and exists in many different forms in the body. These include messenger RNA (mRNA), essential for reading DNA and making **protein**; ribosomal RNA, which is part of the ribosomes, structures that read the mRNA and build the growing **protein**; and a raft of signalling molecules of various sizes and functions, from micro-RNA to long non-coding RNA.

Sequence: This word is used both as a noun — the sequence of a section of DNA is the order of the bases — and as a verb. To 'sequence' a gene is to read its sequence with the goal of either learning what that usually is, or, in medical applications, to compare it with the **reference** to see if there are any medically important **variants**.

Sex chromosomes: The X and Y **chromosomes** are referred to as the sex chromosomes because of their role in determining whether a person is male or female. The most common arrangement is that girls have two copies of the X chromosome and boys have one X and one Y.

Spinal muscular atrophy (SMA): An **autosomal recessive** neurological condition that affects the nerves in the spinal cord that control muscle contraction. SMA varies in its severity — the most common form is lethal in infancy if not treated, but there are later-onset forms as well.

Splicing: The process of removing **introns** from messenger **RNA** as part of the processing needed to make a mature mRNA that can then be translated. Many **genes** can have alternative splicing — some **exons** are variably included or left out — so that the same gene can be responsible for producing multiple different **proteins.**

Telomere: The protective cap at the end of **chromosomes.**

Transcription: The copying of **DNA** into **RNA.** The resulting messenger RNA is processed, including by splicing out the **introns,** to make a mature mRNA. This is then translated to make a **protein.**

Translation: The process of reading messenger **RNA** and translating its message into **protein,** by adding on the encoded **sequence** of amino acids.

Trisomy: literally 'three bodies'. The state of having three copies of a **chromosome** instead of the usual two. Trisomy 21 causes Down syndrome.

Variant: Any difference from the **reference sequence.** This can include variants that have effects on **genes,** and those that are in between genes. Within a gene, there are also different possible consequences from a variant. For example, it might change the **DNA** sequence in an **exon** without changing the resulting **protein** (because the DNA code has redundancy); it might change the sequence so that a different amino acid is substituted for the usual one; or it might introduce a sequence saying 'stop' prematurely. Variants may have different effects on the person in whose **genome** they are found, as well. Many are harmless (Benign). Some are damaging to the gene in a way that can cause disease (Pathogenic). For some, we are not sure of the possible consequences — these are known as Variants of Uncertain Significance. The word 'mutation' technically means the same as variant, but it has long implied that the change is pathogenic; for this reason, the term is falling out of favour. Classification of variants — to decide whether they are relevant to the reason why a test was done — is one of the major challenges in modern genetics, to the extent that there is talk of 'the $1,000 test with a $10,000 interpretation'.

X-linked: A form of inheritance in which there is a variant in a **gene** on the X **chromosome**. This leads to a specific pattern of inheritance. Typically, males are more severely affected by X-linked conditions; females can be affected, usually less severely, but can also have no symptoms at all. If a man with an X-linked condition has children, all his daughters will inherit his X chromosome (that's why they are girls) and will be carriers (or possibly affected); all of his sons will inherit his Y chromosome (that's why they are boys) and will not be affected, nor will they be able to pass the condition on. There are some X-linked conditions that virtually only affect girls, because the effects on a male with no functioning copy of the gene are so severe that no males affected in this way are born.

Acknowledgements

I owe debts of gratitude to many different people for their roles in making this book possible. The chance to write the book came along in a year that was already set to be the busiest of my life, and writing it has often kept me away from my family. I'm deeply grateful to my wife, Sue, and my children, Seamus, Yasmin, and Finn, for their love and support. Every seven years, Sue — but always with you.

The chain of events that led to the writing of the book began with my friend Denny Mrsnik, who introduced me to my agent, Tara Wynne. Tara and Caitlan Cooper-Trent, both from Curtis Brown, have been wonderful to work with. Tara critiqued early chapters and provided valuable guidance, acting as literary coach as much as agent. She took the book to Scribe, and she and Caitlan have been working tirelessly to find other opportunities for me. I hope this will prove to be the start of a long and fruitful partnership.

I am very grateful to all at Scribe. I'd particularly like to thank Henry Rosenbloom (Scribe's founder and publisher) and his editorial team, for taking a chance on a novice writer. My editor, David Golding, has been brilliant. I have David to thank for the title of the book, and for countless improvements throughout the text. David combines great attention to detail with an ability to keep the whole in mind; it has been a privilege working with him. Laura Thomas designed the wonderful cover — I find myself liking it more and more every time I look at it. Mick Pilkington, design and production manager, played a pivotal role in bringing together all the elements of the book: without his work, the physical object you hold in your hands, if you have a physical copy, would not exist. And without the efforts of Chris Grierson from marketing, and Cora Roberts from the publicity department at Scribe, and their teams, you probably never would have heard of the book at all.

I showed drafts of all or part of the manuscript to a number of people. Denny Mrsnik, Seamus Kirk, Sarah Righetti, and Michael Buckley read the whole thing and made numerous helpful comments; Seamus also gave me the quote that starts chapter 1. Others who read sections of the book and suggested improvements, or who made helpful general comments about the book as a whole, include (in no particular order) Colin Nichols, Alan Ma, Jacqui Russell, Lisa Bristowe, Nigel Laing, Martin Delatycki, David Thorburn, Michelle Farrar, Robert Mitchell, Eileen Forbes, Rachael and Jonathan Casella, and Richard Harvey. Tony Roscioli gave me helpful and timely advice about protecting patient confidentiality. Michael and Colin both found scientific errors and corrected them. I have tried hard to make sure that no others remain in the book, but, if there are errors, the fault is entirely my own and not that of any of those who helped me. Michelle Farrar gave me a helpful tutorial on muscle satellite cells, and Finn Kirk pointed out that extracting DNA from strawberries is easier and works better than extracting it from onions, leading to a change of recipe in chapter 2. Professors Peter Campbell and Philip Jones kindly provided the image from their paper used in chapter 3.

At the same time that I was writing the book, the preparations for starting recruitment of couples for Mackenzie's Mission were underway, and I'd like to take this opportunity to thank the many people who have contributed to the project so far. There are more than 80 people who are investigators contributing to the project's committees, or who have other key roles, and dozens more without whose work we could not hope to succeed. I'm especially grateful to the executive team — my fellow leads, Martin Delatycki and Nigel Laing; Jade Caruana, our program coordinator; and Tiffany Boughtwood, manager of Australian Genomics. The project is being delivered through the infrastructure of Australian Genomics, which is led by Kathryn North — who is also a member of our steering committee. Without Rachael and Jonathan Casella, none of this would have been possible, and Rachael has been a valued contributor to the project in many ways, including as a member of the steering committee. In NSW, I'd like to particularly mention Sarah Righetti, our administrator, without whom 2019 would have completely overwhelmed me; Kirsten Boggs, Lucinda Freeman, and Kristine Barlow-Stewart, our genetic counsellors; and all of the team at the NSW Health Pathology Randwick Genomics Laboratory, but especially Corrina Cliffe, Bianca Rodrigues, Natalia Smietanka, Guus Teunisse, Ying Zhu, Janice Fletcher, Tony Roscioli, and Michael Buckley. Thank you all.

Notes

This section is for additional detail, sources, and references, although referencing is *not* intended to be at the level of a scientific paper.

Preface

Genetics has taken me to some unexpected places: This isn't intended to be a book about me, so I never quite found room to expand on this in the main text.

The room in the basement full of mice was the mouse facility at the University of Sydney's School of Veterinary Science, where I spent a lot of time while I was working on my PhD.

I went to Pakistan while taking part in production of a television program for the Discovery channel. This was not a high watermark for television, so I suggest you *don't* seek it out online. But this did give me the chance to spend time in Lahore, a rare privilege. While there, I had the chance to visit the magnificent Badshahi Mosque, a masterpiece of Mughal architecture and probably the most beautiful building I've ever seen.

At the Badshahi Mosque I was just a tourist, but I visited a mosque in Western Sydney for quite a different reason. The latter visit was part of preparations for a study of carrier screening in couples who are related to one another (mentioned in chapter 11). There are many parts of the world where cousin marriages (mainly, although in some places uncle–niece marriages also occur) are culturally favoured, including much of the Middle East. While the practice is not specifically linked to religion — cousin marriages are common among Lebanese Christians, for instance — Islam predominates in those regions, and it was important to us to consult with relevant religious and other community leaders before starting the study. My friend and collaborator Kristine Barlow-Stewart and I visited one large mosque together, among other meetings we had in preparation for the study. I visited another without Kris. At that mosque, I met an imam who had *two* PhDs — possibly the most highly educated person I've ever met, and a kind and attentive host. Having done one PhD myself, I found the idea of someone voluntarily undertaking a second a little hard to grapple with. After our meeting, I was offered a tour of the mosque. The young man in a hoodie who was tasked with showing me around greeted me with a hopeful expression. 'Are you a convert?' he asked.

1. Easier than you think

DNA is a chemical: Okay, here's some chemical detail, if you have a stomach for that sort of thing. The four nucleobases, or bases, that make up DNA are adenine, cytosine, guanine, and thymine (A, C, G, T). DNA is short for 'deoxyribonucleic acid'. Each base of DNA is either a carbon ring (C, T) or double ring (A, G) and is attached to a sugar called deoxyribose and a phosphate; together, the base plus sugar plus phosphate are referred to as a nucleotide. Each nucleotide joins from sugar to phosphate in a long chain; the double helix forms when the rings and double rings of the bases are attracted to each other by hydrogen bonds. C sticks to G with three bonds; A sticks to T with two — so where there are lots of Cs and Gs, the double helix is tighter and harder to separate. There's another important base, uracil, that stands in for thymine in RNA (ribonucleic acid — which has a slightly different sugar backbone). That sounds complicated but it's all just detail. 'DNA is a chemical that contains information ... written in an alphabet with only four letters: A, C, G, T' covers the main points.

the language of DNA has only 21 words: All right — there is just a *little* more to it than that. For a start, there are actually 64 ways of spelling those 21 words (four possibilities for the first letter, times four for the second, times four for the third). This means that there are alternative ways of spelling most of the words — as if kat were a routinely used alternative way of spelling cat. For nine of the amino acids, there are two possible DNA codes; for one, there are three; for eight, there are four. There are just two amino acids — methionine and tryptophan — with unique DNA spellings. There are three ways of coding for 'stop' — TAA, TAG, and TGA all mean this. ATG spells methionine, but it also codes for 'start'. And there is a 21st word, selenocysteine. The amino acid cysteine contains a sulphur, but, in selenocysteine, this is replaced by a selenium. If there's enough selenium about (i.e. you aren't deficient in it), then TGA can mean 'put a selenocysteine here' instead of 'stop'. There are about 50 human proteins that contain selenocysteine, a tiny but important fraction of the whole. Selenocysteine was only discovered in the 1970s, long after the other amino acids, by the American biochemist Thressa Stadtman.

Honestly, though, all of the above is also just detail. None of it really adds much *conceptually* to the text in chapter 1.

that race was won by Lap-Chee Tsui: This was a fiercely competitive international race. Lap-Chee Tsui was working in Toronto, and the work was done in collaboration with Francis Collins, from the US, among others. Tsui, with Ruslan Dorfman, wrote an article called 'The Cystic Fibrosis Gene: a molecular genetic perspective', which describes the discovery of the gene and then goes into considerable detail about its structure — kind of 'everything you wanted to know about *CFTR* (the gene) but were afraid to ask'. It's ... a tad technical, but if you're keen, it's freely available at: www.ncbi.nlm.

nih.gov/pmc/articles/PMC3552342/

Edwards syndrome: Reported in the journal *The Lancet*. Edwards, J.H. et al. 'A New Trisomic Syndrome'. Published as *The Lancet* 1960;1:787–90, but now *The Lancet* 1960;275:787–90. *The Lancet*, which was founded in 1823, has a rather confusing history of changes of volume numbering.

2. The DNA Dinner

announcing that the human genome had been sequenced: There are some marvellous documents about the Human Genome Project on the website of the Oak Ridge National Laboratory (www.ornl. gov) — an entire archive of material about the HGP. For example, there is a transcript of the press conference at the White House on 25 June 2000 at: web.ornl.gov/sci/techresources/Human_Genome/ project/clinton1.shtml

It's well worth visiting and having a poke around.

good working draft: International Human Genome Sequencing Consortium. 'Initial Sequencing and Analysis of the Human Genome'. *Nature* 2001;409:860–921

Venter, J.C. et al. 'The Sequence of the Human Genome'. *Science* 2001;291:1,304–51

still 341 gaps: International Human Genome Sequencing Consortium. 'Finishing the Euchromatic Sequence of the Human Genome'. *Nature* 2004;431:931–45

the genome browser run by the University of California, Santa Cruz: This is available at: genome.ucsc.edu

The European version of this is the Ensembl Genome Browser, at www.ensembl.org. While I mainly use the UCSC browser, that's only a personal preference, and both genome browsers are wonderful. They are freely available to anyone with an interest.

the first scientific report of treatment with insulin: Banting, F.G. et al. 'Pancreatic Extracts in the Treatment of Diabetes Mellitus'. *Canadian Medical Association Journal* 1922;12:141–6

3. The boy who wasn't short

Philadelphia chromosome: This and Rowley's and Garson's discoveries are described in various places, but the latter part can be found in the remarkably detailed history of cytogenetics at St Vincent's Hospital in Melbourne: stvincentsmedicalalumni.org. au/wp/wp-content/uploads/2017/11/2010-Egan-prize-joint-winner_History-of-Cytogenetics-at-SVHM.pdf

as many non-human as human cells in your body: This is controversial, and there are widely varying estimates. I've gone with some work published a few years ago: Sender R., Fuchs S., and Milo R. 'Revised Estimates for the Number of Human and Bacteria Cells in the Body'. *PLoS Biology* 2016;14(8):e1002533 — freely available at: doi.org/10.1371/journal.pbio.1002533. But they may be wrong.

ten quadrillion cell divisions: This one probably comes under the category of 'well-informed wild guess'. There's a source for this at 'the database of useful biological numbers', which is hosted by Harvard University (but could still be very wrong on this point): bionumbers. hms.harvard.edu/bionumber.aspx?s=n&v=10&id=100379

40 to 80 changes in your DNA ... you didn't get from the genome of either parent: This at least is based on high-quality data: Gómez-Romero, L. et al. 'Precise Detection of De Novo Single Nucleotide Variants in Human Genomes'. *PNAS* 2018;115(21):5,516–21

a recent study suggests that, typically, there is one new mistake per cell division: Milholland, B. et al. 'Differences Between Germline and Somatic Mutation Rates in Humans and Mice'. *Nature*

Communications 2017;8:15,183

what they actually found was … horrifying: The source of the image and accompanying information is: Martincorena, I. et al. 'High Burden and Pervasive Positive Selection of Somatic Mutations in Normal Human Skin'. *Science* 2015;348(6,237):880–6

a group of researchers led by Mary-Claire King … narrowed down the location of BRCA1: Hall, J.M. et al. 'Linkage of Early-Onset Familial Breast Cancer to Chromosome 17q21'. *Science* 1990:250(4,988):1,684–9

In May 1994, a group from the University of Utah, in collaboration with a group from Cambridge University in the UK: Albertson, H.M. et al. 'A Physical Map and Candidate Genes in the *BRCA1* Region on Chromosome 17q12–21'. *Nature Genetics* 1994;7:472–9

publication of the sequence of the gene [BRCA1]: Miki, Y. et al. 'A Strong Candidate for the Breast and Ovarian Cancer Susceptibility Gene *BRCA1*'. *Science* 1994;266(5,182):66–71

another British-led group, headed by Michael Stratton, published the sequence of BRCA2: Wooster, R. et al. 'Identification of the Breast Cancer Susceptibility Gene *BRCA2*'. *Nature* 1995;378:789–92

Myriad … had applied for a patent: There is a long but fascinating article about the history of this topic: Gold, E.R. and Carbone, J. 'Myriad Genetics: in the eye of the policy storm'. *Genetics in Medicine* 2010;12(4 Suppl.):S39–S70

The article is freely available online via Pubmed Central and is well worth a read: www.ncbi.nlm.nih.gov/pmc/articles/ PMC3037261/

4. Uncertainty

a majority of couples request termination: A review of 19 studies that looked at choices made after prenatal diagnosis of a sex chromosome condition found that, for Turner syndrome (45,X), pregnancies were terminated on average 76 per cent of the time; for XXY, it was 61 per cent; and for XXX and XYY, 32 per cent. The studies were from 13 different countries, mostly developed nations. Jeon, K.C., Chen, L-S, and Goodson, P. 'Decision to Abort After a Prenatal Diagnosis of Sex Chromosome Abnormality'. *Genetics in Medicine* 2012;14:27–38

5. Needles in stacks of needles

Knome's US$24,500 exome service: MacArthur, D. 'Knome Offers Sequencing of All of Your Protein-Coding Genes for $24,500'. *Wired* 10 May 2009. Available at: www.wired.com/2009/05/knome-offers-sequencing-of-all-of-your-protein-coding-genes-for-24500/

Dan Stoicescu: Harmon, A. 'Gene Map Becomes a Luxury Item'. *The New York Times* 4 March 2008. Available at: www.nytimes.com/2008/03/04/health/research/04geno.html

James Watson's genome: The article describing this is open-access — freely available online — from *Nature*: Wheeler, D.A. et al. 'The Complete Genome of an Individual by Massively Parallel DNA Sequencing'. *Nature* 2008;452:872–6. The senior author on this paper was Jonathan Rothberg. It is available at: www.nature.com/articles/nature06884

studies in fish came to a pretty similar conclusion: Halligan, D.L. and Keightley, P.D. 'How Many Lethal Alleles?' *Trends Genet* 2003;19(2):57–9

a Danish group, led by Morten Olesen: Refsgaard, L. et al. 'High Prevalence of Genetic Variants Previously Associated with LQT Syndrome in New Exome Data'. *European Journal of Human Genetics* 2012;20:905–8

Andreasen, C. et al. 'New Population-Based Exome Data are Questioning the Pathogenicity of Previously Cardiomyopathy-Associated Genetic Variants'. *European Journal of Human Genetics* 2013;21:918–28

Genome In A Bottle: Information about the GIAB consortium is at: www.nist.gov/programs-projects/genome-bottle

For example, the genes CACNB2 and KCNQ1 are both commonly included in panels of genes for testing people with hypertrophic cardiomyopathy: And they aren't the only ones. There's a really useful paper that looks at a long list of genes included in hypertrophic cardiomyopathy panels and concludes that many of them have limited or no evidence for a link to the condition: Ingles, G. et al. 'Evaluating the Clinical Validity of Hypertrophic Cardiomyopathy Genes'. *Circulation: Genomic and Precision Medicine* 2019;12:e002460. The paper is freely available online at: www.ahajournals.org/doi/10.1161/CIRCGEN.119.002460

6. Power!

the story of life on Earth: There's a nice article about this: Marshall, M. 'Timeline: the evolution of life'. *New Scientist* 14 July 2009. Available at: www.newscientist.com/article/dn17453-timeline-the-evolution-of-life/

The earliest person whose name we know: Robert Krulwich of *Radiolab* fame wrote about this: Krulwich, J. 'Who's the First Person in History Whose Name We Know?' *National Geographic* 19 August 2015. Available at: www.nationalgeographic.com/science/phenomena/2015/08/19/

whos-the-first-person-in-history-whose-name-we-know/

nothing but human beings for two kilometres: This assumes a generation time of 33 years, which is conservative, and emergence of modern humans about 200,000 years ago. You can find various estimates for both of these figures. If you assume a generation time of 25 years and emergence of modern humans 300,000 years ago, it would be nothing but humans for four kilometres.

mitochondria live their own little lives: This is fairly full-on science, but, if you are keen, makes for fascinating reading: Sasaki, T. et al. 'Live Imaging Reveals the Dynamics and Regulation of Mitochondrial Nucleoids During the Cell Cycle in Fucci2-HeLa Cells'. *Scientific Reports* 2017;7:11,257

as we age, our mitochondria accumulate damage: see Bua, E. et al. 'Mitochondrial DNA-Deletion Mutations Accumulate Intracellularly to Detrimental Levels in Aged Human Skeletal Muscle Fibers'. *American Journal of Human Genetics* 2006;79(3):469–80

mitochondrial bottleneck: Khrapko, K. 'Two Ways to Make a mtDNA Bottleneck'. *Nature Genetics* 2008;40(2):134–5 — this article is freely available at Pubmed Central: www.ncbi.nlm.nih.gov/pmc/articles/PMC3717270/

a patient who had been seen at Sydney Children's Hospital more than 20 years previously: Lim, S.C. et al. 'Mutations in *LYRM4*, Encoding Iron-Sulfur Cluster Biogenesis Factor ISD11, Cause Deficiency of Multiple Respiratory Chain Complexes'. *Human Molecular Genetics* 2013;22(22):4,460–73

Pauline's story: I've seen families who have experienced similar events, but the specific mutation-load numbers in this story are adapted from a family reported by David Thorburn and colleagues: Thorburn, D.R., Wilton L., and Stock-Myer, S. 'Healthy Baby Girl Born Following Pre-Implantation Genetic Diagnosis for

Mitochondrial DNA m.8993T>G Mutation'. *Molecular Genetics and Metabolism* 2009;98:5–6

7. Dysmorphology Club

Later art contained ever clearer depictions of specific conditions: See, for instance, Bukvic, N. and Elling, J.W. 'Genetics in Art and Art in Genetics'. *Gene* 2015;555(1):14–22

immigration delay disease: Burger, B. et al. 'The Immigration Delay Disease: adermatoglyphia-inherited absence of epidermal ridges'. *Journal of the American Academy of Dermatology* 2011;64:974–80

CATCH22 ... John Burn: Burn, J. 'Closing Time for CATCH22'. *Journal of Medical Genetics* 1999;36:737–8

Eva Sedláčková: Vrtička, K. 'Present-Day Importance of the Velocardiofacial Syndrome'. *Folia Phoniatrica et Logopaedica* 2007;59:141–6

Jacqueline Noonan: Opitz, J. 'The Noonan Syndrome'. *American Journal of Medical Genetics* 1985;21:515–18. This is the same John Opitz after whom numerous syndromes are named.

Bettex and Graf: Bettex, M. et al. 'Oro-Palatal Dysplasia Bettex-Graf — a New Syndrome'. *European Journal of Pediatric Surgery* 1998;8(1):4–8

Julius Hallervorden: Shevell, M. 'Racial Hygiene, Active Euthanasia, and Julius Hallervorden'. *Neurology* 1992;42:2,214–19

John Langdon Down: Down, J. 'Observations on an Ethnic Classification of Idiots'. *London Hospital Reports* 1866;3:259–62

the term 'mongolism' for Down syndrome was common at least

until the 1960s, and probably later: When I was a medical student in the 1980s, I remember the doctors who taught us paediatrics telling us not to use the term, which suggests it may have remained in use by some even then.

Strickland's rules: Rookmaaker, L.C. 'The Early Endeavours by Hugh Edwin Strickland to Establish a Code for Zoological Nomenclature in 1842–1843'. *Bulletin of Zoological Nomenclature* 2011;68(1):29–40

Cantú and his group were expressing uncertainty about this: Garcia-Cruz, D. et al. 'Congenital Hypertrichosis, Osteochondrodysplasia, and Cardiomegaly: further delineation of a new genetic syndrome'. *American Journal of Medical Genetics* 1997;69:138–51

Kathy Grange ... made the inspired observation: Grange, D.K. et al. 'Cantú Syndrome in a Woman and Her Two Daughters: further confirmation of autosomal dominant inheritance and review of the cardiac manifestations'. *American Journal of Medical Genetics A* 2006:140(5):1,673–80

8. How to make a baby

He Jiankui had used the CRISPR technology ... to change the genetic make-up of two babies: BBC News has reported extensively on these events. See, for example, 'China Jails "Gene-Edited Babies" Scientist for Three Years', 30 December 2019, available at: www.bbc.com/news/world-asia-china-50944461

There is a fascinating article about these events: Lovell-Badge, R. 'CRISPR Babies: a view from the centre of the storm'. *Development* 2019;146:dev175778 — freely available at: dev.biologists.org/content/develop/146/3/dev175778.full.pdf

9. Complexity

childproof lid: See the Canadian Medical Hall of fame: www.cdnmedhall.org/inductees/henribreault

Kelsey was a remarkable woman: Bren, L. 'Frances Oldham Kelsey: FDA medical reviewer leaves her mark on history'. *FDA Consumer* March–April 2001. Available at: permanent.access.gpo.gov/lps1609/www.fda.gov/fdac/features/2001/201_kelsey.html

'Heroine' of FDA Keeps Bad Drug Off of Market: *The Washington Post* 15 July 1962

There have been questions raised about how reliably they can assess risks: The work of Karavani, E. et al. (*Cell* 2019;179(6):P1424–1435.E8) provides a detailed critique of the approach. A version of this paper posted ahead of publication is freely available at bioRxiv at: www.biorxiv.org/content/10.1101/626846v1.full

The published research on which the test apparently rests: Lello, L. et al. 'Genomic Prediction of 16 Complex Disease Risks Including Heart Attack, Diabetes, Breast and Prostate Cancer'. *Scientific Reports* 2019;9:15,286

10. A spoonful of mannose-6-phosphate

Jesse Gelsinger: There's an informative piece about the events that led to Jesse's death, and the aftermath, on the website of the Science History Institute. Rinde, M. 'The Death of Jesse Gelsinger, 20 Years Later'. 4 June 2019. Available at: www.sciencehistory.org/distillations/the-death-of-jesse-gelsinger-20-years-later

Was the Human Genome Project a Dud? Torrey, E.F. *Dallas Morning News* 13 October 2019

enzyme replacement therapy: There's an interesting article about Roscoe Brady's work from the Office of NIH History at: history. nih.gov/exhibits/gaucher/docs/page_04.html

11. Please, screen me

tributes written by colleagues after his death are carefully worded: These can be found at: www.robertguthriepku.org/tributes/

resisted the addition of some other conditions to newborn screening: Guthrie, R. 'The Origin of Newborn Screening'. *Screening* 1992;1:5–15

first successful treatment of a child with PKU: Bickel, H., Gerrard, J., and Hickmans, E.M. 'Influence of Phenylalanine Intake on Phenylketonuria'. *The Lancet* 1953;265(6,790):812–13

In 1968, a 29-year-old woman: Valenti, C., Schutta, E.J., and Kehaty, T. 'Prenatal Diagnosis of Down's syndrome'. *The Lancet* 1968;2:220

Malaria remains a serious killer: These are from the World Health Organization www.who.int/news-room/fact-sheets/detail/malaria and the Global Burden of Disease Study — Roth, G.A. et al. 'Global, Regional, and National Age-Sex-Specific Mortality for 282 Causes of Death in 195 Countries and Territories, 1980–2017: a systematic analysis for the Global Burden of Disease Study 2017'. *The Lancet* 2018;392(10,159):1,736–88

Israel is the undisputed world leader: Zlotogora, J. 'The Israeli National Population Program of Genetic Carrier Screening for Reproductive Purposes. How Should It Be Continued?' *Israel Journal of Health Policy Research* 2019;8:73

George Stamatoyannopoulos: Srivastava, A. et al. 'A Tribute

to George Stamatoyannopoulos'. *Human Gene Therapy* 2016;27(4):280–6